CATHEDRAL, FORGE, AND WATERWHEEL

By Frances and Joseph Gies

Life in a Medieval Village (1990)
Marriage and the Family in the Middle Ages (1987)
Women in the Middle Ages (1978)
The Ingenious Yankees (1976)
Life in a Medieval Castle (1974)
Merchants and Moneymen (1972)
Leonard of Pisa *(juvenile)* (1969)
Life in a Medieval City (1969)

Also by Frances Gies

The Knight in History (1984)
Joan of Arc (1981)

Also by Joseph Gies

By the Sweat of Thy Brow: Work in the Western World
   (with Melvin Kranzberg) (1975)
Wonders of the Modern World (1966)
Bridges and Men (1963)
Adventure Underground (1962)

CATHEDRAL, FORGE, AND WATERWHEEL

TECHNOLOGY AND INVENTION
IN THE MIDDLE AGES

FRANCES & JOSEPH GIES

 HarperPerennial
A Division of HarperCollinsPublishers

First HarperPerennial edition published 1995.

Designed by Alma Hochhauser Orenstein

The Library of Congress has catalogued the hardcover edition as follows:
Gies, Frances.
        Cathedral, forge, and waterwheel: technology and invention in the Middle Ages / by Frances and Joseph Gies. — 1st ed.
    p.   cm.
        Includes bibliographical references and index.
        ISBN 0-06-016590-1
        1. Technology—History. 2. Inventions—History. I. Gies, Joseph. II. Title.
    T17.G54   1994
    609.4'09'02—dc20   93-14293

ISBN 0-06-092581-7 (pbk.)

18 19 20 LSC 20 19 18

In memory of Albert Mayio

ACKNOWLEDGMENTS

This book was researched at the Harlan Hatcher Graduate Library of the University of Michigan.

Professor Bert S. Hall of the Institute for the History and Philosophy of Science and Technology of the University of Toronto read the manuscript and provided corrections, improvements, and valuable suggestions.

CONTENTS

1. Nimrod's Tower, Noah's Ark   1
2. The Triumphs and Failures of Ancient Technology   17
3. The Not So Dark Ages: A.D. 500–900   39
4. The Asian Connection   82
5. The Technology of the Commercial Revolution:
   900–1200   105
6. The High Middle Ages: 1200–1400   166
7. Leonardo and Columbus: The End of the Middle Ages   237

   Notes   293

   Bibliography   329

   Index   345

NIMROD'S TOWER, NOAH'S ARK

I N THE CENTURIES FOLLOWING THE MIDDLE Ages, thinkers of the European Enlightenment looked back on the previous period as a time "quiet as the dark of the night,"[1] when the world slumbered and man's history came to "a full stop."[2] A spirit of otherworldliness and a preoccupation with theology were perceived as underlying a vast medieval inertia. The most influential spokesman for this point of view was historian Edward Gibbon, who in his *Decline and Fall of the Roman Empire* described medieval society as "the triumph of barbarism and religion."[3]

Images of lethargy and stagnation were persistently applied to the Middle Ages well into the twentieth century. Even today the popular impression remains to a great extent that of a millennium of darkness, a thousand years when "nothing happened." To the average educated person, the most surprising news about medieval technology may be the fact that there was any.

Yet not all intellectuals of the past shared the negative view of the Middle Ages. In 1550 Italian physician and mathematician Jerome Cardan wrote that the magnetic compass, printing, and gunpowder were three inventions to which "the whole of antiquity has nothing equal to show."[4] A generation later, the Dutch scholar Johannes Stradanus (Jan van der Straet, 1528–1605) in his book *Nova reperta* listed nine great discoveries, all products of the Middle Ages.[5]

Gibbon's eighteenth-century contemporary Anne-Robert-Jacques Turgot, finance minister to Louis XVI, looked back on the Middle Ages as a time when "kings were without authority, nobles without constraint, peoples enslaved ... commerce and communication cut off," when the barbarian invasions had "put out the fire of reason," but he saw it also as a time when a number of inventions unknown to the Greeks and Romans had been somehow produced. Turgot credited the medieval achievement to a "succession of physical experiments" undertaken by unknown individual geniuses who worked in isolation, surrounded by a sea of darkness.[6]

Today, on the contrary, the innovative technology of the Middle Ages appears as the silent contribution of many hands and minds working together. The most momentous changes are now understood not as single, explicit inventions but as gradual, imperceptible revolutions—in agriculture, in water and wind power, in building construction, in textile manufacture, in communications, in metallurgy, in weaponry—taking place through incremental improvements, large or small, in tools, techniques, and the organization of work. This new view is part of a broader change in historical theory that has come to perceive technological innovation in all ages as primarily a social process rather than a disconnected series of individual initiatives.

In the course of recent decades, the very expression "Dark Ages" has fallen into disrepute among historians. The 1934 Webster's asserted that "the term *Dark Ages* is applied to the whole, or more often to the earlier part of the [medieval] period, because of its intellectual stagnation." The 1966 Random House dictionary agreed, defining "Dark Ages" as "1. The period in European history from about A.D. 476 to about 1000; 2. The whole of the Middle Ages, from about A.D. 476 to the Renaissance," a description repeated verbatim in its 1987 edition. The HarperCollins dictionary of 1991, however, recognized the term's decline in scholarly favor, defining "Dark Ages" as "1. The period from about the late 5th century A.D. to about 1000 A.D., once considered an unenlightened period; 2. (occasionally) the whole medieval period."

Recently, historians have suggested the possibility of a narrower use of the old term. In a presidential address to the Medieval Academy of America in 1984, Fred C. Robinson recommended keeping "Dark Age," in the singular, and restricting its meaning to our dim perception of the period (owing to the scarcity of documentary evidence) rather than to its alleged "intellectual stagnation."[7]

The problem of definition also involves the dating of the Middle Ages. The once sovereign date of A.D. 476 as starting point has been judged essentially meaningless, since it marks only the formal abdication of the last Western Roman emperor. In fact, the now general employment of the round A.D. 500 is an admission by historians that there is really no valid starting point, that the beginning of the Middle Ages overlaps and intermingles with the decline and fall of the Western Roman Empire. At the other end, the precise but even less meaningful 1453 (the fall of Constantinople and the end of the Hundred Years War) has been widely replaced by the round 1500, suggestive principally of the opening of the Age of Exploration and the historic impingement of Europe upon America and Asia.

From the third decade of the present century, a recognition of medieval technological and scientific progress has been affirmed by scholars such as Marc Bloch, Lynn White, Robert S. Lopez, Bertrand Gille, Georges Duby, and Jacques Le Goff. Most modern textbooks include in their history of invention the medieval discovery or adoption of the heavy plow, animal harness, open-field agriculture, the castle, water-powered machinery, the putting-out system, Gothic architecture, Hindu-Arabic numerals, double-entry bookkeeping, the blast furnace, the compass, eyeglasses, the lateen sail, clockwork, firearms, and movable type.

But while the pioneering work in medieval technology by Marc Bloch and Lynn White was undertaken in an era (roughly

1925 to 1960) that affirmed human progress and regarded advances in technology as self-evidently positive, the climate of the last part of the twentieth century has become less favorable to technology in general and even to the idea of progress. Suddenly, instead of being credited with no technology, the Middle Ages is found by some to have had too much. Activities once universally regarded as beneficent (such as the land-clearance campaigns of the great monasteries) have been condemned: "The deforestation of Europe during the twelfth century—especially during the 1170s and 1180s—may be seen as the first great ecological disaster," wrote George Ovitt, Jr., in 1987.[8]

Such present-minded thinking permeated Jean Gimpel's *The Medieval Machine* (1976). Drawing a parallel with twentieth-century industrial society, which he envisioned in Spenglerian decline, Gimpel pictured an overindustrialized late medieval Europe suffering from overpopulation, pollution, economic instability, dwindling energy sources, and general malaise.[9]

But despite the many medieval contributions to technology, to speak as Gimpel does of an "industrial revolution" of the Middle Ages is hyperbole. By the same token, pollution was slight, energy sources were largely untapped, the financial crisis of the fourteenth century was temporary and local, and population was excessive only in respect to the limitations of existing agricultural technology. Advanced though it was over the classical age, medieval technology was still in what Lewis Mumford called the "eotechnic" phase—the age of wood, stone, wind, and water—to be followed, in Mumford's terminology, by a "paleotechnic" phase in which coal and iron dominated, and finally by our present "neotechnic" phase of electricity, electronics, nuclear energy, alloys, plastics, and synthetics.[10]

When Gibbon indicted the Middle Ages as "the triumph of barbarism and religion," he coupled the two great bugbears of the intellectual elite of his day, both widely regarded as hostile to scientific and technical progress. The Catholic Church long stood condemned as the enemy of enlightenment, with the alleged suppressions of Copernicus and Galileo as Exhibit A. More recent historians, however, have pointed to evidence of

Church attitudes and policies of a quite different coloration. Lynn White asserted that Christian theology actually gave the Middle Ages a fiat for technology: "Man shares in great measure God's transcendence of nature. Christianity, in absolute contrast to ancient paganism and Asia's religions . . . not only established a dualism of man and nature but also insisted that it is God's will that man exploit nature for his proper ends."[11]

Even earlier, Max Weber (1864–1920) drew attention to the prominent role given by the Benedictine Rule to monastic labor ("Idleness is the enemy of the soul. Therefore the brothers should have a specified period for manual labor as well as for prayerful reading.") and to the well-organized physical self-sufficiency of the monastic community.[12] In the same vein, Ernest Benz pointed to medieval iconography showing God as a master mason, measuring out the universe with compasses and T square, and noted that such images, drawing a parallel between God's labors and those of men, offer an indication of the status of technology in medieval Christendom.[13]

God as master mason measures the universe. [Osterreichische Nationalbibliothek. Codex 2.554, f. 1.]

More recently, George Ovitt, studying the attitudes of medieval theologians, has found that they advocated steward-ship of nature at the same time that ecological evidence shows "an ethic of appropriation" and a "social commitment to the primacy of human habitation" over competing interests.[14] Their varying and contradictory attitudes, he has concluded, represent a rationalization "in response to changes in the 'struc-tures of everyday life' that were created by others,"[15] that is, in response to what was actually going on in the real world.

The forces that impelled medieval men to clear land for cul-tivation and to develop new ways of exploiting nature were complex, but they were surely social and economic rather than ethical or religious. And while the monasteries were among the great clearers of land, the chief conservationists of the Middle Ages were the kings and great lords, who stringently protected their forests, not as guardians of nature, but in the interest of the aristocratic recreation of hunting (just as latter-day hunters' organizations help to preserve wilderness).

Did Christian theologians of the Middle Ages believe, as Lynn White wrote, that "it is God's will that man exploit nature for his proper ends"? And were the theologians' attitudes toward labor and the crafts as benign as Ernest Benz thought?

One of the early Church Fathers, Tertullian (c. A.D. 160–240), commented eloquently on the effects of human enterprise on the earth: "Farms have replaced wastelands, culti-vated land has subdued the forests, cattle have put to flight the wild beast, barren lands have become fertile, rocks have become soil, swamps have been drained, and the number of cities exceeds the number of poor huts found in former times . . . Everywhere there are people, communities—everywhere there is human life!" To such a point that "the world is full. The elements scarcely suffice us. Our needs press . . . Pesti-lence, famine, wars, [earthquakes] are intended, indeed, as remedies, as prunings, against the growth of the human race."[16]

Tertullian anticipated Malthus in his gloomy view. He was echoed by St. Augustine (A.D. 354–430), who cited Adam's

Fall as the dividing point between man's living in harmony with nature and his exploiting it. Prelapsarian (before the Fall) Adam dwelt peacefully in a world where conception occurred "without the passion of lust," childbirth without "the moanings of the mother in pain," where man's "life was free from want . . . There were food and drink to keep away hunger and thirst and the tree of life to stave off death from senescence . . . Not a sickness assailed him from within, and he feared no harm from without."[17] But where prelapsarian Adam lived wholesomely within nature, postlapsarian Adam lived greedily off its bounty. Only by recovering their moral and spiritual innocence could Adam's successors restore the perfection of the world before the Fall.

In the eighth century Anglo-Saxon theologian and historian Bede expanded on Augustine, picturing Adam and Eve before the Fall as vegetarians, living on fruits and herbs and practicing agriculture as an idyllic pastime, symbolic of the cooperation between human beings and a benign nature. With the Fall, as man turned exploiter, Bede agreed, he lost his natural sovereignty.[18] Five centuries after Bede, at the height of the Middle Ages, St. Thomas Aquinas echoed his and Augustine's message and further rationalized it by asserting that "by the very course of nature . . . the less perfect fall to the use of the more perfect," and therefore man holds power over the animals and the rest of the natural world. Before the Fall, all remained obedient to man, like domestic animals, but man's reign was not exploitative. Adam governed by reason, for the common good. Only with the fall of reason was the providential order overthrown.[19] Thus medieval theologians' interpretation of the Creation and the Fall revealed a God-ordained world dominated by human beings, whose role in respect to nature, however, was not exploitation but stewardship and cooperation.

Commentaries on Adam's Fall illuminated one aspect of the Church's fundamental posture in respect to technology. Another lay in the theologians' attitudes toward labor and toward crafts and craftsmen. Ambivalence was characteristic of both.

Benign Adam names the animals.
[Bodleian Library. Ashmole Bestiary,
Ms. Ashmole 1511, f. 9.]

From its earliest beginnings, Christian monasticism emphasized the importance of labor in the interests of the communal life and of humility. In the religious settlements founded in Egypt by Pachomius (A.D. 290–346), productive labor was treated as beneficial both materially and spiritually. Bishop and chronicler Palladius (A.D. 363–431) reported that at one Pachomian settlement he saw monks working at every kind of craft, including "fifteen tailors, seven metalworkers, four carpenters, twelve camel-drivers, and fifteen fullers."[20] St. Jerome (A.D. c. 347–420) described a similar community: "Brothers of the same craft live in one house under one master. Those, for example, who weave linen are together, and those who weave mats are looked upon as being one family. Tailors, carriage makers, fullers, shoemakers—all are governed by their own masters." Labor was accompanied by spiritual exercises and discipline.[21]

The Benedictine Rule, composed in the sixth century, similarly mingled labor and prayer. Labor supported the community, discouraged idleness, and taught obedience and humility; its ends were primarily spiritual. In the following centuries, monasticism struggled to keep the balance between spirituality and economic self-sufficiency. In the course of time, Benedictine monasteries became the victims of their own success, as they grew wealthy from rents, church revenues, gifts, tithes, and other fees, and labor ceased to be performed by the monks but was delegated to peasants and servants.

Late in the eleventh century, the new Cistercian Order attempted to return to the letter of the Benedictine Rule. Founding communities in the wilderness, far from centers of population, the Cistercians divested themselves of many of the sources of income exploited by the Benedictines and at the same time tried to restore the model of manual labor performed by the community itself. The order's outstanding leader, Bernard of Clairvaux (1090–1153), believed that work and contemplation must be kept in balance. The ideal monk was one who mastered "all the skills and jobs of the peasants"—carpentry, masonry, gardening, and weaving—as a means of bringing order to the world.[22]

The Cistercians, however, soon attempted to solve the problem of balance by splitting St. Bernard's ideal monk in two, assigning prayer and work to different categories of brothers, drawn from different social classes. Alongside the regular monks, with aristocratic backgrounds, they established an order of lay brothers, *conversi*, recruited from the lower classes, to perform their communities' skilled labor, supplemented by hired unskilled laborers. Like their predecessors, the Cistercians grew rich, and as the numbers of *conversi* declined and communities relied increasingly on hired labor, they found themselves following the very practices that they had renounced.

Much the same fate befell similar efforts by other monastic orders, and the exemplar of the Benedictine Rule, the monk who prayed, labored with his hands, and studied the Bible, was abandoned.[23] Where the Cistercian pioneer Aelred of Rievaulx (c. 1110–1167) "did not spare the soft skin of his hands," according to his biographer, "but manfully wielded with his slender fingers the rough tools of his field tasks,"[24] his contemporary, Premonstratensian monk Adam of Dryburgh, expressed feelings shared by many fellow monastics in complaining that "manual labor irritates me greatly" and declaring that agricultural work should be performed not by educated and ordained men but by peasants accustomed to hard labor.[25] Work was no longer an integral part of the service of God.

An element in the failure to incorporate labor successfully into the monastic life on a permanent basis was the fact that, as Europe's new intellectual class, the churchmen were the inheritors of a long tradition of disdain for what Aristotle called the "banausic," or utilitarian arts, "the industries that earn wages," that "degrade the mind" and were unworthy of the free man.[26] These arts might have practical value, Aristotle conceded, but "to dwell long upon them would be in poor taste."[27] Aristotle's prejudice was sustained by most of the Greek and Roman philosophers and thinkers. Even Cicero, who extolled man's ability to change his environment through technology, thought that "no workshop can have anything liberal about it."[28]

The Church Fathers retained some of the classical attitude but showed a new interest in and enthusiasm for what St. Augustine called "our human nature" and its "power of inventing, learning, and applying all such arts" as minister to life's necessities and "to human enjoyment." Augustine pointed to "the progress and perfection which human skill has reached in the astonishing achievement of clothmaking, architecture, agriculture, and navigation . . . in ceramics . . . drugs and appliances . . . condiments and sauces." These accomplishments were the products of "human genius," but, he added, this genius was often used for purposes that were "superfluous, perilous, and pernicious." What was most needed, in Augustine's eyes, was a capacity for "living in virtue" in the grace of God.[29]

Boethius, the last great Roman intellectual (c. 480–524),

followed the classical tradition in devising an educational curriculum composed of the seven liberal arts, organized into the *trivium* (grammar, rhetoric, and logic) and the *quadrivium* (arithmetic, geometry, astronomy, and music), with no room for the vulgar "banausic" arts frowned on by Aristotle. Boethius's elitist classification became the basis of the medieval educational system, but other contemporary writers included the crafts at least as secondary adjuncts. The Greek historian Cassiodorus (c. 490–c. 585) wrote enthusiastically about inventions used in the monastery that he had founded: the "cleverly built lamps," the sundials and water clocks, the water-powered mills and the irrigation system, the Egyptian-invented papyrus—"the snowy entrails of a green herb, which keeps the sweet harvest of the mind, and restores it to the reader whenever he chooses to consult it." Mechanics, he judged, was a wonderful art, "almost Nature's comrade, opening her secrets, changing her manifestations, sporting with miracles."[30]

A century later encyclopedist Isidore of Seville showed a lively interest in technology, devoting six of the twenty books of his *Etymologies* to the vocabulary of crafts—the types and elements of ships, buildings, clothing, weapons, harnesses, and household utensils—and classifying mechanics, astrology, and medicine as elements of physics and philosophy.[31] A major innovation in semantics followed in the era of Charlemagne when philosopher John Scotus Erigena invented the term *artes mechanicae* (mechanical arts), describing these as supplements of the liberal arts.[32] Following his lead a tardy three centuries later, scholar Honorius of Autun described ten liberal arts: the usual seven, plus *physica* (medicine), economics, and *mechanica*: "Concerning mechanics . . . it teaches . . . every work in metals, wood, or marble, in addition to painting, sculpture, and all arts which are done with the hands. By this art Nimrod erected his tower, Solomon constructed his temple. By it Noah fashioned his ark, and all the fortifications in the entire world were built, and it taught the various weavings of garments."[33]

Finally, the economic revival of the high Middle Ages, accompanied as it was by a flood of technical advances, stimulated the exploration of new ways of integrating technology into the circle of human knowledge, usually as a physical and material side of theoretical science. The most comprehensive of these systems was that of Hugh of St. Victor (1096–1141), a German theologian who taught in Paris and who compiled an encyclopedic work called the *Didascalicon*. Hugh advocated a life of contemplation that, however, included secular learning. In correspondence with the seven liberal arts of Boethius, he envisioned seven categories of mechanical arts: textile manufacture, armament, navigation, agriculture, hunting, medicine, and theatrics, describing each in detail, for example:

> Textile manufacture includes all types of weaving, sewing, and spinning which are done by hand, needle, spindle, awl, reel, comb, loom, crisper, iron, or any other instrument out of any material of flax or wool, or any sort of skin, whether scraped or hairy, also out of hemp, or cork, or rushes, hair, tufts, or anything of the kind which can be used for making clothes, coverings, drapery, blankets, saddles, carpets, curtains, napkins, felts, strings, nets, ropes; out of straw, too, from which men usually make their hats and baskets. All these studies pertain to textile manufacture.[34]

In Hugh's classification, the mechanical art of *navigatio* (navigation) included not only the techniques of sailing but commerce itself—"every sort of dealing in the purchase, sale, and exchange of domestic or foreign goods." Hugh extolled all those who courageously penetrated "the secret places of the world," approaching "shores unseen" and exploring "fearful wildernesses," bringing peace and reconciliation to all nations and commuting "the private good of individuals into the common benefit of all." His "armament" was a similarly broad category, including architecture, carpentry, and metallurgy; "hunting" included food gathering, cooking, and selling and serving food and drink; "theatrics" all kinds of games and amusements.[35]

Although ranking technology lowest among the arts, Hugh accorded it a moral value, conferred by God as a partial remedy for man's fallen condition. Other creatures were born clothed: "Bark encircles the tree, feathers cover the bird, scales encase the fish, fleece clothes the sheep, hair garbs cattle and wild beasts, a shell protects the tortoise, and ivory makes the elephant unafraid of spears." Only man "is brought forth naked and unarmed"; therefore man was equipped with reason, to invent the things naturally given to the other animals. "Want is what has devised all that you see most excellent in the occupations of men. From this the infinite varieties of painting, weaving, carving, and founding have arisen, so that we look with wonder not at nature alone but at the artificer as well."[36]

Other twelfth-century thinkers adopted Hugh's classification, accepting technology as a part of human life, inferior to intellectual and spiritual elements but necessary and natural. Technology made life easier, freeing the mind from material concerns and supplementing man's innate powers. Hugh's influence extended to such thirteenth-century luminaries as Albertus Magnus, St. Bonaventure, Vincent of Beauvais, and Robert Kilwardby. In his *De ortu scientiarum* (On the origin of sciences) Kilwardby (d. 1279) dignified the mechanical arts by explaining them as practical divisions of the speculative sciences. Every speculative science had a practical aspect. Science explained the *propter quid*, the reason for being, the cause; the mechanical arts the *quia sunt*, the way things are. The two lent each other mutual support. Geometry was necessary to carpenters and masons, astronomy to navigation and agriculture. Wool manufacture was subject to mathematics, since it "examines the number and texture of threads and the measurement and form of the warp, stating in each of these matters that it is this way or that way, while mathematics examines the causes. Similarly all other mechanical arts are found to be under some speculative science or sciences."[37]

Kilwardby, in a word, replaced Hugh's concept of the moral value of technology with that of an intellectual value, a more modern view but one that subordinated technology to the theoretical sciences. The English Franciscan Roger Bacon (c. 1220–1292) carried the relationship a daring step further, awarding precedence to technology; in Bacon's eyes the practical arts gave man a power over the natural world that theoretical science could never provide. Practical science, he speculated, had almost unlimited application and, like all other knowledge, was given to man "by one God, to one world, for one purpose," as an aid to faith and remedy for the ills of the world.[38]

Thus the Church's attitude toward technology, evolving from diverse sources over time—Adam's Fall, the monastic experience, the classifications of knowledge—may be described as ambivalent, but on balance positive.

How large was technology's role in the social changes that took place during the thousand years of the Middle Ages? Pioneer historian of technology Richard Lefebvre des Noëttes, writing in the 1930s, saw the adoption of a new harness that transformed the horse into an important draft animal as the chief factor in the decline and near disappearance of slavery.[39] In 1940 and subsequently Lynn White broadened Lefebvre's thesis with the bold assertion that the dominant social and political systems of the Middle Ages owed their origins to technological innovations: feudalism to the stirrup, the manorial system to the heavy wheeled plow.[40] Such radical determinism could not fail to provoke a flood of research and analysis by other scholars, and we now acknowledge that, as usual, the truth is far from simple. Technology is only one of the forces, along with new social and economic patterns, that formed medieval society, but for a long time, until Lefebvre and White, it was the most neglected.

Human history records a number of technological "revolutions," the first to be pointed out and labeled being the Industrial Revolution of the eighteenth century. The first to take place, however, was the invention of tools, in effect, the discovery of technology itself, a determining factor in the distinction between man and the other animals. The second, what anthropologist Gordon Childe called the "Neolithic Revolution," was the shift from hunting and gathering to cultivation.

A third was the creation of the great irrigation civilizations of Mesopotamia, Egypt, the Indus Valley, and China, which generated cities, governments, and most of our institutions.[41]

Today we recognize that one of the great technological revolutions took place during the medieval millennium with the disappearance of mass slavery, the shift to water- and windpower, the introduction of the open-field system of agriculture, and the importation, adaptation, or invention of an array of devices, from the wheelbarrow to double-entry bookkeeping, climaxed by those two avatars of modern Western civilization, firearms and printing.

A historical surprise uncovered by recent scholarship, especially through the work of Joseph Needham and his colleagues at Cambridge University, is the size and scope of technology transmission from East to West. Scores of major and minor inventions were introduced from China and India, often through the medium of Islamic North Africa and the Near East. The channels of transmission to Europe are sometimes easy to trace, or to postulate, sometimes more mysterious. Technology traveled with merchants on their trade routes, both overland and by sea; it moved with nomads, armies, and migrating populations; it was carried by ambassadors and visiting scholars, and by craftsmen imported from one country to another. Sometimes the transmission was direct and total. Sometimes, as Needham proposes, "a simple hint, a faint suggestion of an idea, might be sufficient to set off a train of development which would lead to roughly similar phenomena in later ages, apparently wholly independent in origin . . . Or the news that some technical process had successfully been accomplished in some far-away part of the world might encourage certain people to solve the problem anew entirely in their own way."[42]

Armed with innovative technology, both borrowed and homegrown, the European civilization that Edward Gibbon believed had been brought to a long standstill by "the triumph of barbarism and religion" had in reality taken an immense stride forward. The Romans so congenial to Gibbon would have marveled at what the millennium following their own era had wrought. More perceptive than Gibbon was English scientist Joseph Glanvill, who wrote in 1661: "These last Ages have shewn us what Antiquity never saw; no, not in a dream."[43]

Technology—Aristotle's "banausic arts"—embraces the whole range of human activities involving tools, machines, instrumentation, power, and organization of work. What follows in this book cannot attempt to be, even in a compact or shorthand sense, a complete history of Western technology from A.D. 500 to A.D. 1500. Its intention is limited to the identification of the main technological elements that entered significantly into medieval European history, their known or probable sources, and their principal impacts.

2

THE TRIUMPHS AND FAILURES OF ANCIENT TECHNOLOGY

NEARLY EVERYTHING THAT SIXTH-CENTURY Europe knew about technology came to it from Rome. Rome, however, invented few of the tools and processes it bequeathed to the Middle Ages. Roman civilization achieved a high level of culture and sophistication and left many monuments, but most of its technology was inherited from the Stone, Bronze, and early Iron Ages.

From the long Paleolithic (Old Stone) Age came the tools and techniques that separated humankind forever from the animal world: language, fire making, hunting weapons and methods, domestication of animals. From the short Neolithic (New Stone) Age, beginning about 8000 B.C. in Mesopotamia, came agriculture and its tools—plow, sickle, ax, and mortar and pestle or stone grain crusher. The wheel and axle appeared in Mesopotamia between 3000 and 4000 B.C. The arts of cloth making were invented: felting, matting fibers together by boiling and beating to produce a nonwoven fabric; spinning, drawing out fibers of flax or wool and twisting them into a continuous strand, usually by means of a spindle; weaving, interlacing threads with the aid of a loom; fulling, soaking and beating cloth to remove grease; and dyeing. Raw hides were converted into leather by scraping and soaking with tannin, derived from oak bark. The important art of pottery making first modeled clay with fingers and thumb, then coiled strands of clay, and finally shaped its work with the potter's wheel, invented about 3000 B.C.

Copper, sometimes found in a free metallic state, was used by Neolithic man as a substitute for stone, wood, and bone long before the addition of a small amount of tin, probably by accident (c. 3500 B.C.), created the superior alloy bronze. The brief Bronze Age that followed overlapped the Neolithic Age at one end and the longer (still going on) Iron Age at the other. The two metal ages constitute not so much historical periods as stages in technological evolution that took place over different times in different places. The Bronze Age never occurred in pre-Columbian America, where accessible tin was lacking. In the Near East copper continued to be widely used, but the harder yet malleable bronze made better tools and especially better weapons, including the arms and armor of Homer's heroes. Besides its hardness, bronze had a low melting point that permitted casting in molds.

As the Bronze Age introduced "the first great technical civilizations" (Bertrand Gille),[1] the long, unrecorded life of the Stone Ages gave way to written history (including much written in the archaeological record). Civilized communities grew up in widely separated places, with little contact, or no contact at all, with each other. To the Roman and early medieval European worlds, societies in Africa, southeast Asia, Oceania, and America remained totally invisible. Even China and India, whose civilizations rivaled or surpassed those of the West, were scarcely glimpsed across the barrier of geographical distance. Only the civilizations that grew up on the banks of the Tigris-Euphrates and the Nile connected closely with their successor Greco-Roman societies and so contributed significantly to the Roman legacy to medieval Europe.

Besides inventing writing (in the form of the ideograph), the peoples of Mesopotamia (Sumerians, Babylonians, Assyrians) and the Egyptians of the Nile pioneered astronomy, mathematics, and engineering. Their river-dependent agriculture inspired the first dams and canals, and the first water-lifting device, the *shaduf* or swape (c. 3000 B.C.), a counterweighted lever with a bucket on one end. Cultivation of grape and olive stimulated the invention about 1500 B.C. of the beam press, worked by a lever. Fermentation, discovered by the Egyptians, converted grape juice into wine and cereal into bread or beer; the rotary quern, invented about 1000 B.C., speeded the universal daily labor of milling. Techniques of food preservation—drying, salting, smoking—were invented (or more likely discovered). Cloth makers invented the vertical loom described by Homer, the "great loom standing in the hall" with "the fine warp of some vast fabric on it," in Penelope's artfully unfinished task.[2] Cities built the first water-supply and drainage systems; street paving was pioneered in Babylon and road paving in Crete.[3] Egypt and Babylon produced the first clock to supplement the ancient sundial: the clepsydra, or water clock, a vessel out of which water ran slowly, with graduated marks to indicate the passage of hours as it emptied. It operated at first with mediocre accuracy, since as the water diminished the flow slackened.[4]

Like bronze, iron came on the scene by accident. Because iron has a higher melting point than copper, it could not easily be separated from its ore but had to be hammered loose. Even then it found little use for a thousand years after its first discovery (c. 2500 B.C.), until smiths in the Armenian mountains near the Black Sea found that repeated heatings and hammerings in a charcoal fire hardened it.[5] In the *Iliad*, weapons are made of bronze, tools of iron, "the democratic metal."[6]

The irrigation civilizations of the Nile and Tigris-Euphrates built temples, palaces, obelisks, and tombs, the Egyptians of the early dynasties (third millennium B.C.) employing copper tools, ramps, levers, and guy ropes, but neither pulley nor wheel. The massive blocks of stone that formed the Pyramids were hauled on boards greased with animal fat and raised to the upper courses by means of earthen ramps, afterward removed. While the Mesopotamians made some use of the arch to support their roofs, Egypt and Greece relied on the post and lintel (two vertical columns joined at the top by a horizontal member). Pericles' Athens borrowed Egyptian stonemasonry techniques, such as the assembling of columns out of stacks of drums, while strengthening their structures with metal strips, pins, and clamps. The beams that held up the ceiling of the Propylaea on the Acropolis (440–430 B.C.) were reinforced with iron bars, the first use of metal structural members in building construction. Mesopotamia, poor in wood and stone, invented brick making, first with sun-dried brick in Sumer (before 3000 B.C.), later with kiln-dried brick in Babylon.[7]

The horse was tamed by at least the eleventh century B.C.,[8] but the absence of saddle and stirrups limited its military value, while the problem of harness reduced its role as a draft animal. The throat-and-girth harness that suited the configuration of the ox choked the horse, which could consequently pull only light loads, such as the two-wheeled war chariot of the *Iliad*. At the same time, lack of a firm saddle handicapped pack animals.

While land transportation hardly progressed between Neolithic and Roman times, water transportation made a great leap forward. By 1000 B.C. the Phoenicians, the master mariners of the ancient world, were building ships with stempost, sternpost, and skeleton of ribs that reinforced hull planking fitted edge to edge and joined by mortise and tenon—in a word, modern construction.[9] Homer, writing in the seventh or eighth century B.C., depicted Odysseus single-handedly building the boat that carried him from Calypso's isle, boring his timbers with an auger and fastening them together with wooden dowels.[10]

Ships used both sail and oar. The early Egyptians paddled facing forward; the oar, a less obvious device than the paddle, turned the crew around and faced them backward. The sail may also have been born on the Nile, where prevailing winds conveniently blow in the direction opposite to the current; Egyptians sailed up and floated down their great river. The single sail (cotton, linen, or Egyptian papyrus) was square, rigged at right angles to the hull. Steering was done with a large oar mounted on one side near the stern. Navigation was by sun and stars and the unaided eye, and by dead reckoning: a rough calculation of the ship's speed, course, and drift. With such ships and techniques, the Phoenicians ("greedy knaves," according to the *Odyssey*)[11] not only sailed and rowed from their homeland (roughly modern Lebanon) the length and breadth of the Mediterranean but ventured into the Atlantic after British tin.

Needing written records and communications, Phoenician mariner-merchants invented one of the alphabets (as opposed to ideographs) of the ancient world, the one that passed, with variations, to the Greeks, thence to the Romans, and so to medieval Europe. Its spread was assisted by the advent of the second of the world's three great writing materials, parchment, the dried, stretched, and shaved hide of sheep, goats, and calves, smoother and more durable than Egypt's reed-derived papyrus. Parchment received its final improvement in the second century B.C. in Greek Pergamum (whence the name "parchment"), in the form of slaking in lime for several days. Both sides of the resulting material could be written on and the leaves bound into a book (codex), more convenient than the ancient scroll.

Most of the military history of the ancient world is irrelevant to the record of humanity's progress, but the conquests of Alexander the Great in the late fourth century B.C. had the significant effect of promoting the "Hellenization" (Hellas: Greece) of the whole Near East and eastern Mediterranean. The succeeding age is famous for its philosophers, mathematicians, and natural scientists, headed by Alexander's own tutor, Aristotle (384–322 B.C.). Although Aristotle shared the prejudice of his master Plato against the arts and crafts, among the works attributed to him or (more recently) to his pupil Strato is *Mechanics*, the world's first engineering text. *Mechanics* contains the earliest mention of multiple pulleys and gear wheels, along with all the simple mechanical-advantage devices except the screw.

Alexander's eponymous city on the shore of Egypt, Alexandria, came to house the greatest library of learning in the

Mediterranean world and to shelter some of the greatest scientists. These included the mathematician Euclid (fl. c. 300), Eratosthenes (c. 276–194 B.C.), who made the first calculation of the earth's circumference, and the astronomer-geographer Ptolemy (fl. A.D. 127–145). The aim of the dilettante scientists of Hellenistic Greece was "to know, not to do, to understand nature, not to tame her" (M. I. Finley).[12] Nevertheless, they made serious contributions to technology as well as to science. Archimedes (c. 287–212 B.C.) discovered the principle of buoyancy and stated that of the lever. Another of the basic machine components, the screw, has been attributed to him but may have existed earlier: in its original form a water-lifting device, a spiral tube inside an inclined cylinder turned by slaves or animals walking a treadmill. Archimedes may also have invented the toothed wheel and gear train, first described in Western writings by him.[13]

Two other Alexandrians who left evidence of inventive minds and outlooks were Ctesibius (fl. 270 B.C.) and Heron (fl. first century A.D.). Ctesibius discovered the compressibility of air and probably invented the force pump, a pair of cylinders whose pistons were driven by a horizontal bar on a fulcrum between them, alternately forcing the water out of one and drawing it into the other. He also solved the problem of the water clock's irregularity by providing an overflow outlet that kept the water in the operative vessel at constant depth.[14] Heron invented a number of mechanical toys, including a miniature steam engine, creations whose principles would eventually be applied to practical uses but only after the world had passed through several preparatory revolutions.

The Hellenistic Greeks did not invent but gave impetus to the two great "false sciences" of alchemy and astrology, speculative parents of chemistry and astronomy. Both originated in Mesopotamia at a very early date, and both were actively pursued in the Hellenistic age. A late addition to astrological theory, the casting of the individual's horoscope, had valuable consequences for science, since it demanded an accurate knowledge of the motions of the planets to determine their position at the hour of birth.

Hellenistic astrological interest resulted in the anonymous invention at Alexandria of the astrolabe, "the world's first scientific instrument."[15] In its original form, the astrolabe ("astro"-"labe," star-plate) was a wooden disk bearing a map of the heavens, its outer edge marked off in 360 degrees. A pointer pivoted on a central pin could be aimed at the sun or other celestial body to give the altitude above the equator, providing a reasonably accurate indication of the time of day for a given latitude. Conversely, the astrolabe could determine latitude, but no one thought of this possibility for a long time.

Astrology passed from the Greeks to the Romans and thence to medieval Europe, while alchemy, disdained by the Romans, reached medieval Europe only at a later date, via the Arabs. But as Roman conquest absorbed the Hellenistic world, an enormous transfer of technology took place, from the Phoenician-Greek alphabet to Archimedes' screw to masonry construction. Roman technology was strongest where Rome's predecessors were strongest, weakest in areas which they had neglected or where they had failed.

The Romans inherited most of their agricultural tools and techniques, improving and adding to them. The *aratrum*, the light plow that worked satisfactorily in the sandy soils of the Mediterranean region, was made more effective by two additions—first, an iron coulter, a vertical blade fixed in front of the plowshare, and, second, a wooden moldboard behind it to turn the soil. The Romans' engineering approach to agriculture improved irrigation systems and pioneered the systematic application of fertilizer. Although they did little scientific breeding of plants or animals, they increased the numbers of horses and sheep and found a better method of harvesting wool, applying shears in place of the traditional method of plucking during the molting season.[16]

The grinding of grain received a worthwhile Roman improvement in the transformation of the rotary hand quern into the large donkey- or slave-powered hourglass mill, examples of which are preserved in Pompeii, Herculaneum, and Ostia. The processing of grape and olive was likewise improved by the adoption of the screw press, a useful new application of Archimedes' screw with significance for the distant future.[17]

From the Greeks, the Romans received a well-developed mining technology along with the system of operating mines as a government monopoly, relying on slave labor and iron tools: hammer, pick, chisel, wedge. Pillars were left to support headings; niches were cut in the walls to hold oil lamps. Ventilation remained an unsolved problem, conditions of labor miserable.[18] To the iron metallurgy they inherited from the Greeks, the Romans added tempering (reheating and cooling), which hardened the metal without making it brittle. To their inherited tool chest they added the carpenter's plane, which first appears

Roman grain mills in Herculaneum. Grain was poured into an opening in the center of the upper millstone, the flour falling into a trough around the base of the lower stone. A beam inserted through the square holes in the upper millstone served as a handle for turning the stone, either by slaves or donkeys. The mill on the right has lost its upper stone.

*Photographs are the authors' unless otherwise credited.

in Roman representations and may have been a Roman invention.[19]

Handicraft production flourished in the Roman Empire, fostered by larger markets and the growth of an affluent class of city dwellers. The chief industry was the manufacture of wool and linen cloth (Chinese silk and Egyptian cotton were imported luxury fabrics). Women did the spinning and weaving at home or on the great estates, their instruments the ageless spindle and the vertical loom. Finishing—fulling and dyeing—required a capital outlay and therefore passed into the hands of male specialists working in shops.[20]

Roman potters followed the Greek tradition that had carried the craft to artistic heights, but without improvements in processes or materials. Glass manufacture, however, whose techniques lay somewhere between ceramics and metallurgy, achieved a major innovation: glassblowing, invented in the Roman province of Syria in the first century A.D.[21]

Fuller's shop in Pompeii, trough for soaking textiles. Although in antiquity spinning and weaving were domestic industries performed by women, finishing was done by male specialists.

Like Egypt and Greece, the Roman Empire left its most conspicuous achievements in its building construction. Employing engineering technology on a scale never before seen in the Western world, it strewed the Mediterranean littoral and western Europe with bridges, roads, walls, public baths, sewage systems, arenas, forums, markets, triumphal arches, and theaters. Among the most characteristic of Roman ruins are the aqueducts that served the water-supply system of the capital and other cities. Generally they ran in low, open or covered masonry channels or in conduits tunneled through hillsides, but at times they strode across valleys in long, picturesque lines of stone arches. One of the most impressive of Roman relics is the triple-tiered Pont du Gard in southern France, whose two main tiers have stood for two thousand years without the aid of mortar. The Romans possessed an excellent lime mortar but used it only for construction with smaller stones, such as those in the top tier of the Pont du Gard. By mixing their mortar with a sandy volcanic ash, Roman builders produced a hydraulic cement, one that dried to rock hardness underwater. Mixed with sand and gravel, it became waterproof concrete.[22]

The basic design component of Roman construction was the semicircular arch, converted by extension into the barrel vault, capable of carrying a greater load and spanning a greater breadth than a simple beam. With this strong, enduring, and versatile device the Romans built aqueducts, bridges, baths, and basilicas that stood for centuries. Yet there was a blind spot in the Roman dependence on the semicircular form. As a vault, it placed tremendous weight on the supporting walls, which had to be made thick and nearly windowless. As an arch in a bridge, it required massive piers in the stream, mounted on the always uncertain base of sapling poles driven in the river bottom to "refusal," that is, as deep as men standing in the water and mud could drive them. Cofferdams (temporary watertight enclosures built in the stream) permitted deeper-driven piles, but the resulting piers remained vulnerable to scour, the abrasive action of the current swirling sand

The Pont du Gard, Roman aqueduct spanning the Gard River.

around the pier footings. Scour was itself heightened by the constriction imposed on the current by the many thick piers. Though a number of Roman bridges endured, many fell victim to scour.[23]

Roman engineering, which learned surveying from the Egyptians, stressed exact measurements and imposed on the Western world the system of weights and measures (inch, foot, mile, pound, amphora) that the Greeks had adapted from the Egyptians, Phoenicians, and Babylonians. Besides their monumental public works, the Romans created fine domestic architecture for their wealthy class, by far the largest and richest of the classical world. In the multistoried houses of the crowded capital, they introduced the interior stairway, while in the roomier countryside they built the comfortable and aesthetically pleasing one-story villa, home to provincial government officials and well-to-do private families. From the Roman public baths, the villa borrowed its heating system, the India-originated hypocaust, which circulated hot air under a tile floor.[24]

The Ponte Sant' Angelo, Rome. Semicircular arches required massive piers in the stream. [Philip Gendreau.]

One of the most admired Roman engineering works was the vast road network, begun under the Republic and by the third century A.D. comprising 44,000 miles of thickly layered, well-drained, durable roadway, grouted with concrete and topped with gravel, or, in the vicinity of cities, surfaced with flagstones laid in mortar. Typically the road ran straight as an arrow, favoring ridges over valleys and accepting steep grades rather than deviating from the most direct route. Tunneling through rock was done only when unavoidable, employing the Greek method of heating the rock face by building a bonfire, then cracking it by splashing water against it, a technique not improved on until the introduction of explosives.[25]

The preference for straight over level in roads reflected the priority of military use—marching men—over commercial—wagons and pack animals. Land transport remained difficult and expensive, the cost even rising in the late Empire, handicapping economic development.

Shipping by sea was far cheaper, even though few innova-

Paved street in Pompeii.

tions in shipbuilding or navigation were introduced. A long-standing division of ships into two types, "long" and "round," gained sharp definition. Long ships (galleys) were oar propelled, had little cargo space in their narrow hulls, and were employed mainly for war. Round ships were sail powered, deep hulled, clumsy to maneuver, but strong and comparatively durable. Roman shipbuilders followed the Greeks and Phoenicians in laying their planks edge to edge and in building the shell first, inserting the skeleton of ribs afterward, and securing the mortise-and-tenon joints by wooden pegs held by iron nails, making seams so watertight that no caulking was needed. The steering oar was retained, more firmly secured by a boxlike structure that functioned like an oarlock.[26]

The largest navigational problem came in tacking against the wind, which involved sailing a series of zigzags while taking the wind at an angle to the ship's course. A valuable aid of undetermined origin appeared in the Mediterranean as early as the first century A.D. in the form of the lateen sail, a triangular fore-and-aft sail capable of taking the wind on either surface. Shifting it, however, was a difficult task, made more difficult by increasing size, and throughout the Roman era the lateen appeared only on small craft.[27]

Manmade harbor works had been pioneered by the Greeks in the mole at Delos of the eighth century B.C. Roman construction technology multiplied port facilities and lighthouses (copied

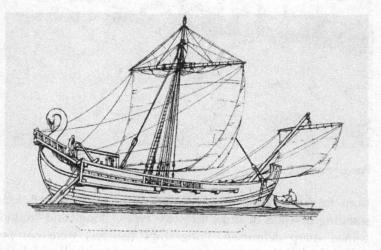

Roman merchant ship, square sailed, deep hulled, maneuvered by steering oar. [Science Museum, London.]

from the famous Pharos of Alexandria) all around the Mediterranean and up the Atlantic coast, where sturdy Roman masonry structures kept beacon fires burning into the Middle Ages.

Notwithstanding their impressive military history, the Romans were not very innovative in equipping their armed forces. The thirty-plus legions who manned the defense perimeter of the vast Empire wore and carried more metal than any army ever had before, but neither arms nor armor offered anything new. The legions' siege artillery was the torsion-powered catapult long used by the Greeks. Its commonest form employed a pair of springs made of bundles of animal sinew, stretched tight and given a twist, to supply power to a giant bowstring.[28] Otherwise the Romans generally disdained the bow, sometimes to their disadvantage. In war as in building construction, organization was the Romans' strong suit. Their echeloned table of organization—legion, cohort, and century—continued unmatched as a command-control system until modern times. So did the legions' unrivaled engineering capability, permitting swift construction of camps, fortifications, roads, and bridges.

Not quite all the technology of the Roman Empire was drawn from the ancient Egyptians, the Near East, and the Greeks. From Gaul in the fourth century A.D. came a long-needed improvement in the processing of the harvest, the jointed flail, created by hinging two sticks together to produce a threshing device much handier than a single stick or the tramp of animals' hooves.[29] Gaulish agriculture also invented an astonishing piece of farm machinery, a mechanical harvester, described by Pliny (A.D. 23–79) as "an enormous box with teeth, supported on two wheels." The machine was still in use in the fourth century A.D., when Palladius left a description that much later, in the 1830s, inspired "Ridley's stripper," an Australian invention.[30] The original harvester disappeared in the early twilight of the Middle Ages. The Gauls were also the source of a form of soap made from fats boiled with natural soda (Romans did not use soap).[31]

Other borrowed technology came from the "barbarians," the epithet under which the Romans (like Gibbon) lumped the immigrants from the north and east who entered the Empire in various ways, peaceable or otherwise, starting in the second century A.D. Though the Germanic intruders lacked such southern refinements as written language and masonry construction, they brought to the Roman world several important innovations including, surprisingly enough, a better grade of metal for weapons. By hardening the surfaces of several thin strips of iron, then welding a bundle of them together, their smiths could achieve an exceptionally hard and durable blade. The operation was chancy, however, and such layered "steel" weapons were costly rarities.[32]

The Germanic peoples also introduced a non-Mediterranean style of clothing that included furs, stockings, trousers, and laced boots, along with the idea of sewing a garment together from a number of separate pieces—in short, modern Western-style clothing and manufacturing technique.[33] Another barbarian contribution, the wooden barrel, began by the first century B.C. to replace fragile clay amphorae and leaky animal skins for transporting oil, wine, and beer.[34]

Despite their engineering skills and talent for creative borrowing, the Romans were technologically handicapped by two momentous failures in the exploitation of power. The first was the shortcoming of the horse harness, unimproved since the Bronze Age. In China, by at least the second century B.C., horses were pulling against a breast strap that allowed them to breathe freely, while the presence there of the even more efficient collar harness was attested pictorially a century later.[35] Yet the Greeks and Romans hit upon neither device. Harnessing in tandem, turning sharply, suspension, and lubrication provided subsidiary problems in vehicular transportation. "The ancient harness . . . enlisted only in feeble measure the strength of each animal, foiling collective effort, and consequently providing only a trifling output" (Lefebvre des Noëttes).[36]

The second failure was in the exploitation of an invention of capital importance, the waterwheel. The Romans did not overlook the waterwheel entirely, but they failed to realize its potential.

The early history of this invention—or inventions, the vertical and horizontal wheels probably having separate origins—is obscure and controversial. The horizontal waterwheel, now believed to have originated in the mountains of Armenia about 200 B.C., seems to have developed directly from the rotary quern. It consisted of a paddle-armed wheel either laid horizontally in the stream with one side masked against the current or furnished with a chute to guide the flow. Suited to streams with a small volume of water and moderate current, it could be readily harnessed to a grain mill by extending the vertical axle upward to a rotating millstone. Simple and cheap to build, it diffused rapidly.[37]

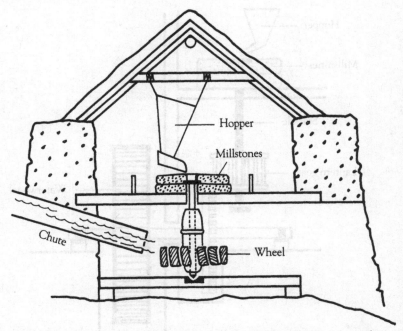

Mill powered by horizontal waterwheel. A chute delivers water to one side of wheel.

The more high-powered vertical wheel evidently derived from a water-lifting device called the "noria," invented in either Persia or India. In its original form, the noria was a large vertical wheel, its circumference armed with buckets, that was turned by oxen circling a capstan or walking a treadmill.[38] But when the noria was mounted in a rapidly flowing stream, the current sufficed to turn the wheel, suggesting the possibility of using it to grind grain. The horizontal axle was extended to turn a pair of gear wheels at right angles to each other, the second of which was made to turn a millstone set above or below it.

The first description of a waterwheel that can be definitely identified as vertical is that of Vitruvius, an engineer of the Augustan Age (31 B.C.–A.D. 14), who composed a ten-volume treatise on all aspects of Roman engineering. Vitruvius

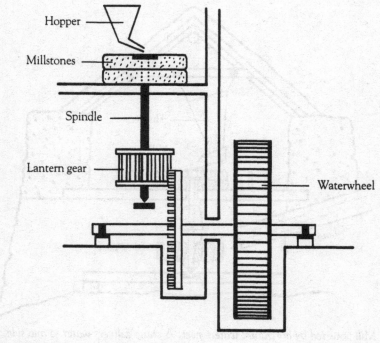

Mill powered by vertical waterwheel.

expressed enthusiasm for the device but remarked that it was among "machines which are rarely employed."[39] The wheel he described was "undershot," that is, the lower part was immersed in the stream so that the current turned it in a reverse direction.

The undershot wheel typically achieved an efficiency of 15 to 30 percent, adequate for milling. For more demanding tasks, a superior design was the overshot wheel. In this arrangement the stream was channeled by a millrace or chute to the top of the wheel, bringing the full weight of the water to bear, with a resulting efficiency of 50 to 70 percent.[40] Because it required dam, millrace, sluice gates, and tailrace as well as gearing, the overshot wheel had a high initial cost. Consequently, large landowners and even the Roman state were reluctant to build it. Few water-powered mills of any type were built outside the cities, though a remarkable complex at Barbegal, near Arles, in southern France, has been identified from ruins. Dating from the fourth century A.D., it consisted of eight overshot wheels, each turning a pair of millstones, with a total capacity of three tons of grain per hour. A tantalizing reference to a waterwheel employed to cut and polish marble also dates from the fourth century, in a passage of the Gallo-Roman poet Ausonius (c. 310–c. 395). This is the solitary reference in any text to a Roman application of waterpower for a purpose other than grinding grain, and its authenticity has been questioned.[41]

What may be said with assurance is that water mills remained scarce in the late Roman Empire, vertical wheels scarcer, the more efficient overshot type scarcer yet, and non-milling applications barely, if at all, existent. To the Empire's end the two great power sources were men and animals, and the animal power was severely handicapped by the want of a good horse harness.

Besides these two technological failures, the Romans may be found guilty of two failures in other realms that exercised large influence on technology: theoretical science and economics. In science, where the Greek elite favored knowing over doing, the

Roman educated class did the opposite, emphasizing doing at the expense of knowing. They took so little interest in Greek science and philosophy that they never bothered to translate Aristotle, Euclid, Archimedes, and other Greek savants into Latin. The consequence was that the intellectual class of medieval Europe, inheriting Latin as its lingua franca, for six centuries remained unaware, or hardly aware, that the Greek classics existed—perhaps the strangest hiatus in the history of Western culture.

The eclipse of Greek learning was not quite total. A few Roman writers, such as Pliny and Boethius, knew their Aristotle. Some, too, made their own original scientific contributions. Out of his personal experience, Columella (fl. first century A.D.) supplied a guide to scientific farming, *De re rustica* (On rural management), while Vitruvius, the architect-engineer, drew on both his own firsthand knowledge and Greek sources in his massive work. But for the most part theoretical science was underemployed by the Romans in dealing with technical problems. One explanation that has been offered blames the rhetoric-based Roman education system, which in emphasizing composition, grammar, and logical expression rather than knowledge of nature reflected what Lynn White called "the anti-technological attitudes of the ruling class."[42] An outstanding product of that system, the philosopher Seneca (4 B.C.–A.D. 65) seemed to sense the Roman shortcoming when he wrote, "The day will come when posterity will be amazed that we remained ignorant of things that will to them seem so plain."[43]

The final Roman weakness bearing on the history of technology was in the realm of economics. The imposing political and military facade of Imperial Rome masked a chronically impoverished and largely stagnant peasant economy. The great landowners, who relied on slave gangs—whipped, branded, and shackled—to work their plantations (latifundia),[44] had little incentive to explore labor-saving technology, nor were their slaves potential customers who might stimulate investment of capital in enterprises such as grist mills.

While the Imperial government grew to dimensions dwarfing anything seen previously, at least in the West, the Roman private economic sector remained stunted. The Mediterranean port cities sustained an active commercial life, but the scale was small and the business technology primitive, lacking credit instruments, negotiable paper, and long-term partnerships. The only capital resource available on a large scale belonged to the government, which spent generously on roads, public buildings, water supply, and other civic amenities but contributed little to industrial and agricultural production. Private wealth was either squandered on consumption or immobilized in land rather than invested in enterprise.[45]

The Roman economy, in short, was weak in the dynamics that make for the creation and diffusion of technological innovation. The succeeding age, developing different social and economic structures, created a new environment more congenial to technology.

EUROPE, A.D. 500

The fundamental processes of agriculture, pottery making, and cloth making, plus language, fire making, tools, and the wheel, all came out of the Stone Ages, before recorded history began. Metallurgy, writing, mathematics, astronomy, engineering, grape and olive cultivation, food preservation, shipbuilding, and cities were products of the early historic civilizations that flourished in the Near East and Egypt (and in China and India) long before Greece and Rome came on the scene. The two great classical societies in fact "together added little to the world's store of technical knowledge and equipment," as M. I. Finley has noted, citing "a handful of specifics," including gears, the screw, the screw press, glassblowing, concrete, the torsion catapult, automata, and the invention but scanty diffusion of the waterwheel, "not very much for a great civilization over fifteen hundred years."[46]

Nevertheless, Greece and Rome improved on much of the technology they borrowed, and Rome vastly expanded its application. Borrowing technology is a highly worthwhile activity, often leading to further advances that the lending civilization fails to achieve. The new Europe that succeeded the Roman Empire profited from Rome's assiduous borrowing and synthesizing, and launched its own career of doing much the same.

3

THE NOT SO DARK AGES

A.D. 500-900

THE ROMAN EMPIRE, WITH ITS WIDE GEO-graphical extent, sophisticated political and military organization, and stately monuments, made a powerful impression on generations of later historians, who were correspondingly appalled by its collapse. The immediate cause seemed obvious. "The Roman world," wrote Gibbon, "was overwhelmed by a deluge of barbarians."[1] Italy, Gaul, Spain, Britain were overrun by assorted Goths, Huns, Vandals, Franks, Burgundians, Lombards, and Anglo-Saxons, driven west and south by forces that are still unexplained. The Western Roman Empire, long sovereign over the Mediterranean basin, was shattered into fragments governed by these "barbarians." Several generations of scholars debated the sources of the weakness that permitted the calamity. But in the twentieth century, largely owing to the pioneering work of Belgian scholar Henri Pirenne (1862–1935), historians began to shift the sense of their question. Instead of asking what caused the fall of Rome, they began to ask, What exactly *was* the fall of Rome?

Primarily, the fall of Rome was a political event, the disappearance, or radical alteration, of a governmental system. Even in the political sphere it was limited geographically to the western half of the Roman Empire, leaving the eastern (Byzantine) half, with its capital of Constantinople, to survive another thousand years. The fall was also limited in scope, the new local rulers retaining much of the Roman administrative apparatus. Pirenne employed the metaphor of an ancient palazzo that was not razed but subdivided into apartments. Not until the rise of the Arab empire in the seventh century, Pirenne believed, did the classical world collapse, commerce and urban life dwindle, and the Roman administrative framework disappear.[2] The Pirenne thesis stirred controversy and revision, ending with a consensus among scholars, aided by recent archaeology, to the effect that a general social and economic decline took place, later than historians had previously believed, but before the Arabs arrived on the scene.

What actually fell in the "fall of Rome"? In the realm of technology, very little. Lynn White went so far as to assert that there was "no evidence of a break in the continuity of technological development following the decline of the Western Roman Empire."[3] In some regions, certain Roman craft skills were lost for a time. The potter's wheel disappeared from Britain, but when it returned from continental Europe in the ninth century, it had been improved by the kick wheel, which allowed the potter to use both hands to manipulate the workpiece.[4] Roman mining operations contracted under the late Empire, and their scale was not again reached until at least the central Middle Ages, but techniques were not lost, and by the eighth century new mining regions in central and eastern Europe were beginning to open up. Along with mining, metallurgy went into a late Roman decline but by the ninth century showed an upward trend.[5] Similarly, Roman irrigation works in Spain and Africa were lost through neglect in the wake of invasion and war.

Lynn White made his case a little too strongly when he asserted, "In technology, at least, the Dark Ages mark a steady and uninterrupted advance over the Roman Empire."[6] Nevertheless, early medieval technical innovations had an unquestionable impact, helping to bring about Europe's first great transformation. Most of the innovations applied to agriculture, and most were borrowed. In the words of Carlo Cipolla, "What the Europeans showed from the sixth to the eleventh centuries was not so much inventive ingenuity as a remarkable capacity for assimilation. They knew how to take good ideas where they found them and how to apply them on a large scale to productive activity."[7]

The post-Roman world was divided geographically not only between Byzantine East and barbarian West but even more meaningfully between rich South and poor North. The Mediterranean littoral, though the scene of a good deal of political and military turbulence, remained in the late fifth century populous and productive, dotted with cities, towns, and landed estates. To the north, also, little was changed—a sparse population dwelling in temporary farming settlements, few cities worth the name, much empty forest, heath, and swamp. The population density of Gaul in the sixth century has been estimated at 5.5 per square kilometer, that of Germany and Britain at 2.2 and 2.0 respectively. The scattered inhabitants of these cold lands evidently did not live well; their skeletons indicate malnutrition. Famine, plague, and typhus were probably even more endemic here than in the South.[8] Yet this northern region had important natural assets: abundant forests, fast-growing vegetation, accessible metal ores, and numerous rivers and streams, many swift flowing and ice free, with potential beyond transportation and communication. Like the steady winds in other regions of the North, they promised energy sources of immense value. "They were to the people of the time what coal, oil, and uranium are to an industrialized society" (Carlo Cipolla).[9]

While the round-number dates of A.D. 500 to 1500 are now widely accepted for the whole period of the Middle Ages, divisions within the era remain arbitrary. As a terminal date for the early part of the period, the year 900 may be satisfactory, bearing in mind that nothing in particular transpired that year, any more than in the commencement year of 500. As it happens, however, each of the four centuries thus encompassed is marked by its own special catalog of events.

Sixth century, A.D. 500–600: the barbarian century, with the last of the Great Migrations, the establishment of the barbarian kingdoms, and the counteroffensive against the Goths by the Eastern (Byzantine) Roman Empire that turned Italy into a battlefield. "At the end of the sixth century, Europe was a profoundly uncivilized place," Georges Duby observes.[10] There is also reason to believe that the population of the Mediterranean West declined through this century and into the next.

Seventh century, A.D. 600–700: the Muslim century, with the explosion of Islam in North Africa and the Near East. By the end of the century, all the southern regions of the old Roman Empire, plus Persia, were Muslim. Almost overnight a major new power thus appeared, positioned geographically between Europe and Asia-Africa.

Eighth century, A.D. 700–800: the Carolingian century. The first great Carolingian, Charles Martel, halted the Islamic advance into Europe at the battle of Poitiers (or Tours); his grandson Charlemagne founded a short-lived ersatz Roman Empire and promoted the scholarly and artistic revival known as the Carolingian renaissance.

Ninth century, A.D. 800–900: the Viking century, marked by raiding and pillaging of towns and monasteries of western Europe by Scandinavian pirates. Muslim raiders did the same for southern Europe, where the Mediterranean was turned into "a no-man's land between Christian and Muslim naval forces" (Richard Unger).[11]

Most of the violent events that formed the traditional history of the early Middle Ages were, however, essentially superficial. The basic wealth of a peasant economy is land, and land is immune from theft and pillage. Over the four centuries of the early Middle Ages, the value of European land was substantially enhanced by the operation of a demographic phenomenon of much larger effect than all the marauding and pillaging. This was the northward expansion of the population. By the time of Charlemagne, as Pirenne pointed out, the center of gravity of Western civilization had shifted from the Mediterranean to the plains of northern Europe. Coincidentally, and probably somewhat causally, a major meteorological change had occurred. The southward drift of the glacial front that commenced in the fifth century reversed itself in the middle of the eighth. As the frost retreated, northern Europe became more hospitable to agriculture.[12] Scanty data indicate yields per acre well below what the best farmlands of the ancient world produced,[13] but by the seventh century the farming communities of Britain, Gaul, the Low Countries, and Germany were harvesting surpluses sufficient to support a modest but definite population increase.

There was even a little urban growth. In the seventh and eighth centuries, specialized trading settlements called "emporia" or "gateway communities" sprang up near the North Sea and Channel coasts as the Frankish (Merovingian and Carolingian) kings exchanged goods with Anglo-Saxon and Scandinavian chieftains in treaty arrangements called "trade partnerships." The more advanced Frankish rulers offered prestigious commodities such as wines, glassware, and wheel-thrown pottery in return for raw materials like wool and hides, collected as taxes from the chieftains' subjects. By the mid–eighth century emporia such as Hamwih (later Southampton) and Ipswich, in East Anglia, laid out in a grid pattern of workshops, stalls, and storehouses, were among the largest towns in England.[14] London at the time was a "beach-market" (*ripa emptoralis*), serving mostly local traders, farmers, and fishermen, who sold their wares directly from their boats without benefit of docks, shops, warehouses, or middlemen.[15] Across the channel, Dorestad, base of the Frisian traders, and Quentovic, south of Boulogne, flourished until their decline in the ninth century, with the breakup of the Carolingian Empire.[16]

Meanwhile two widespread, technologically related developments changed the face of Europe: a new form of agricultural organization, equipped with a new type of plow; and the rise of a new military caste, composed of armored horsemen, who became for a considerable time the ruling European elite.

A Revolution in Agriculture

The old notion that agriculture stood still or regressed for several centuries in the Middle Ages has long been exploded. Instead, two separate revolutions in the organization of agricultural work took place, one in the early Middle Ages, the other, to be described in chapter 5, in the central Middle Ages.

The first revolution, reinforced by the introduction of two tools that may also be called revolutionary, brought about the disappearance of the old Roman latifundium, slave manned and market oriented. In its place, by the eighth century, stood the estate, equally large but based on a different principle of exploitation: farm labor performed by tenants who divided their time between the lord's land and their own small holdings. Those who were classed as unfree (eventually called "serfs," or in England "villeins") were subject to a varying list of obligations and liabilities not imposed on free tenants. Yet the serfs as well as their free neighbors had a recognized (in the medieval vocabulary a "customary") right to the use of their land, a right, moreover, that was inheritable. Alongside serfdom, slavery persisted but as a marginal and declining institution.[17]

Instead of profit in the marketplace, the new agriculture sought local self-sufficiency and, though falling short of complete success, created a highly decentralized rural landscape. At first its technology was somewhat retrogressive, as such Roman skills and practices as the grafting of fruit trees and application of lime for fertilizer slipped into disuse in many regions. Roman agricultural treatises were neglected and no new ones written. The techniques of cereal-crop production, the main form of agriculture, remained for a time unchanged.

But beginning in the sixth century, a radical improvement in farming's most basic tool was introduced. Pliny had described secondhand a heavy plow, mounted on wheels and drawn by several oxen, reported in use in the eastern provinces of the Roman Empire. Its diffusion must have been limited; in effect, it waited in the wings for five centuries before appearing in numbers sufficient to attract notice, first in the Slavic lands, then in the Po valley, and in the early eighth century in the Rhineland. Sometimes it was mounted on wheels, sometimes not; the main function of the wheels was to allow adjustment of the plowshare to the depth of furrow.

An improved harness for harnessing in tandem (one animal behind the other) facilitated the use of multiple-ox teams to pull the heavy plow in attacking new ground. The combination of plow and team supplied the technological key to the prodigious task of clearing the forestland of fertile northwest Europe. Other new or little-used implements came into wide service: the harrow, which by crumbling the clods after plowing saved laborious cross-plowing; the scythe, rarely employed by the Romans, now needed to cut hay to feed the numerous oxen; and the pitchfork, to handle the hay. When Charlemagne proposed a new nomenclature for the calendar, he renamed July "Haying Month."[18]

Toward the end of the period, an innovation as important for agriculture as the heavy plow made its appearance in Europe: the rigid, padded horse collar, long known in Asia, which converted the horse for the first time into an efficient draft animal. Developed back in Roman times, probably by the horse-dependent nomads of the central Asian steppes, the

Light plow, without coulter or mouldboard, as seen in the Utrecht Psalter (c. 830). [British Library, Harley Ms. 603, f. 54v.]

Heavy plow, with coulter and mouldboard, drawn by four oxen. From the fourteenth-century Luttrell Psalter. [British Library, Ms. Add. 42130, f. 170.]

horse collar progressed westward in a course that has been traced by scholars through linguistic and iconographic clues. The first pictorial evidence of its appearance in Europe occurs in an illumination of the *Trier Apocalypse* (c. 800), which shows a pair of horses pulling an open carriage or wagon. The earliest text reference—to a horse-drawn plow—is from late-ninth-century Norway.[19]

The new device replaced the old throat-and-girth harness which choked the horse when the animal pulled against it. The padded collar, instead of bearing on the trachea, exerted its pressure on the sternum, freeing the respiratory channel and at least tripling the weight a horse could pull. Another practical harness, the breast strap, arrived from China at about the same time but was never widely used in the West.[20]

Faster-gaited and longer-working than the ox, the horse proved under most conditions a superior plow animal and a far better transport beast. The nailed iron horseshoe, also arriving from Asia in the ninth or tenth century, further improved his quality and durability on the farm and on the road. The first pictorial representation of a horse pulling an agricultural implement—a harrow—occurs in the Bayeux Tapestry of circa 1080; by that time the sight was doubtless common.[21]

Yet the ox, the age-old "engine of the peasant," did not retire from the scene. Slow moving but very strong, he had the advantage over the horse in difficult ground, as in first-time

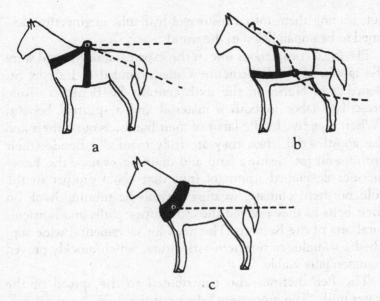

The three main forms of horse harness: (a) throat-and-girth (Western antiquity); (b) breast strap (ancient and early medieval China); and (c) padded horse collar (late medieval China and medieval Europe). [From Joseph Needham, Science and Civilization in China, *Cambridge University Press.]*

plowing of newly cleared land. He was cheaper to feed and in England enjoyed the added appeal of being edible. Pope Gregory III in 732 barred horse meat from the Christian table, an injunction that for unknown reasons was respected only in England. On the Continent, old plow horses were eaten with as much relish as old oxen.[22]

As the new plow, pulled by whichever traction animal, proved its ability to cultivate the rich, heavy soils of northwest Europe, the region's forest, moor, and swamp were attacked with ax and spade. Even the sea was made to contribute new land for cultivation. The inhabitants of the low-lying Netherlands coast built dikes to protect themselves from storms and abnormally high tides; gradually the deposits of silt that collected became new dry land at normal high tide. The Netherlanders appropriated it by building new dikes farther out, leading them into a history of hydraulic engineering destined to be unparalleled in the world.

Playing a conspicuous role in the expansion northward were the monks of the Benedictine Order, founded in Italy by St. Benedict of Nursia in the sixth century. St. Benedict's Rule prescribed labor as both a material and a spiritual benefit: "When they live by the labor of their hands, as our fathers and the apostles did, then they are truly monks."[23] Besides their enthusiasm for clearing land and draining swamp, the Benedictines developed strains of fruit that could prosper in the cold northern climate, wearing a symbolic pruning hook on their belts as they revived the old grafting skills and horticultural arts of the Romans. The need for sacramental wine supplied a stimulus to northern viticulture, which quickly proved commercially viable.

The Benedictines also contributed to the spread of the water mill. "The monastery," declared the Rule, "ought if possible to be so constituted that all things necessary, such as water, a mill, a garden, and the various crafts might be contained within it."[24] Both city and country followed the Benedictine lead in exploiting fast-flowing year-round streams. Gregory of Tours (538–594) described several mills, including one at Dijon, where the river Ouche "turns the millwheels round at wondrous speed outside the gate,"[25] and another on the Indre, constructed by Ursus, abbot of Loches, "made . . . with wooden stakes packed with large stones and sluice-gates to control the flow of water into a channel in which the millwheel turned."[26]

The law code issued by Frankish king Clovis in about 511 imposed fines for stealing grain or iron tools from another man's mill or breaking into his mill enclosure.[27] That of Lombard king Rothair, promulgated in 643, stipulated fines for burning another man's mill, breaking his dam, or building a mill on a neighbor's part of the riverbank.[28] By the time of Charlemagne, mills were important enough to be taxed in the imperial *Capitulare de villis* (800).[29]

Built on many of the great estates of the ninth century, water mills represented a substantial investment but produced lucrative profits in the form of "multure," a percentage of the peasants' grain or flour exacted at the mill.[30] On one manor of the abbey of St.-Germain-des-Prés, millers delivered as much grain in multure from the peasants every year as the lord's own fields produced. The polyptych (estate survey) of the abbey, dated 801–820, lists no fewer than fifty-nine mills, including eight new and two recently renovated.[31]

Hardly anything is known about the configuration of these early medieval mills. The horizontal wheel, needing no gearing, was easy to build and repair, and consequently popular. Wealthy lords, however, may have built the more powerful and efficient vertical wheels.[32]

By the tenth century, the water mill had achieved a status and value far beyond what it had possessed under the Roman Empire. It made a significant contribution to the agricultural revolution wrought by the horse harness, the heavy plow, and the self-contained tenant-farmed estate.

Cloth Making: Women's Work

Agriculture developed a new social and economic function in the early Middle Ages while improving its technical equipment; cloth making retained its equipment while undergoing modest alterations in function. As in Roman times, women dominated manufacture. Their tasks, as indicated by a statute of 789, included not only spinning and weaving but shearing sheep, crushing flax, combing wool, and cutting and sewing garments.[33]

Free women and serfs worked in their homes, slave women in the workshops (gynaecea) of the great estates. Almost every estate of any importance had a gynaeceum. Gregory of Tours mentions "the women who worked in the spinning and weaving room" of the royal manor of Marlenheim.[34] At the council of Nantes in 660, the prelates chided aristocratic women for attending public assemblies and "usurping senatorial authority" when they "ought to be sitting among their girls of the cloth shop and ought to be talking about their wool processing and their textile labors."[35] Several Germanic law codes mention women's workshops, while the Carolingian *Capitulare de villis* prescribed that women in the gynaeceum should be supplied with "linen, wool, woad, red dye, madder, carding implements, combs, soap, oil, containers, and other small things that are needed there."[36]

The workshops were sometimes located in sunken huts, whose earthen floors were excavated two or three feet below ground level, the interiors lighted by an opening in the roof. Alternatively, a two-story building provided sleeping quarters in the upper floor. If the cloth was to be dyed, the shop included a hearth to heat water. In summer the looms might be set up outdoors in open, roofed structures or under canopies.[37]

As in Roman times, linen and wool remained the principal textiles. The manufacture of cotton cloth, a Roman luxury import, was carried by the Arab conquest to Spain, Sicily, and southern Italy as early as the tenth century, but it was not mastered by Christian Europe until the twelfth. Silk, China's most celebrated export, became the object of a historic coup of industrial espionage in the sixth century. Two Greek monks journeying to China are said to have secreted silkworm cocoons in their staffs and returned to Constantinople to launch the Byzantine silk industry. The story, as Joseph Needham has pointed out, leaves puzzling questions: presented with the cocoons, how did the Byzantine textile workers acquire the techniques of unreeling and processing the fibers?[38] Evidently the information was somehow made available, for the Byzantine court soon had its own silk-weaving establishment, in addition to privately owned workshops. Silk manufacture, however, did not penetrate western Europe until the eleventh and twelfth centuries.

Wool cloth, indispensable in the cold climate of newly developed northern Europe, retained its dominant position among textiles, as the center of gravity of the industry moved northward to northern France and the Low Countries. When Harun-al-Rashid, the caliph of the *Thousand and One Nights*, sent Charlemagne gifts that included "many precious silken robes," a linen tent, perfumes, ointments, an elaborate water clock, and an elephant, Charlemagne replied with a present of Spanish horses and mules, hunting dogs, and "some [woolen] cloaks from Frisia, white, gray, crimson, and sapphire-blue"— made of the expensive cloth traded by Frisian seamen. (Harun, unimpressed, "cast a careless eye" over everything but the hunting dogs.)[39] Linen, an article of commerce in the Roman period, in the early Middle Ages retreated to the status of a domestic industry, supplying local needs but no longer profitable enough to transport to distant markets.[40]

Techniques of manufacture remained unchanged over a long period. Wool fleece was given a preliminary washing, then combed to remove the tangles and impurities and draw out the fibers parallel to one another. Yarn was spun with the spindle, usually "suspended"—free hanging—in a process unchanged since its description by Catullus in the first century B.C. Holding in her left hand the distaff, a short forked stick around which a mass of the prepared raw fibers was wound, the spinster took some of the fibers between the finger and thumb of her right hand, twisting them together as she drew them gently downward. When the thread thus produced was long enough, she tucked the distaff under her arm or in her belt and tied the thread with a slipknot to the top of the spindle, a toplike rod with a disk-shaped weight attached to the bottom to increase rotation, and gave it a turn. The suspended weight pulled the fibers slowly through the spinster's fingers, while the rotation twisted them together into yarn. The process depended on the practiced skill of the spinster in controlling the release of the fibers. Drawing out more fibers from the distaff, she repeated the operation until the spindle reached the floor, when she picked it up and wound the spun thread around it. When the spindle was full, she wound the thread into a ball.[41]

The process never ceased, and the skill was universal, especially for women of the lower classes, who always had spindle in hand, even while cooking, feeding livestock, or minding the children (or, to believe one medieval miniature, having sex). Spinning was so identified with women that the female side of the family was known as the "distaff side," or the "spindle side." Primitive though the technology seems, hand spinning created an excellent product, one not easily matched by machinery even centuries later.

It took many hand spinners to supply a single weaver, who operated one of two types of vertical loom, either warp-weighted or two-beam. In its most primitive form, the warp-weighted loom consisted of a pair of wooden uprights joined at the top by a wooden "cloth beam" that could be turned to roll up the cloth as it was woven. Warp (lengthwise) threads, hanging from the cloth beam, were held taut by clay weights at the bottom. To produce a plain weave, the weaver might first pass the weft (lateral) thread from right to left over each even-numbered warp thread and under each odd-numbered one, then on the next row return from left to right, reversing the procedure, lifting the warp threads with one hand as the weft was passed under with the other. A ninth-century saint's life describes a woman weaver in "the winter work halls" of an estate weaving "with bent fingers," holding a small skein or ball of weft in her hand and passing it through the warp.[42]

Distaff and suspended spindle: women carried them even when performing other tasks. [British Library, Luttrell Psalter, Ms. Add. 42130, f. 166v.]

After a row was completed, the weft was pushed up to join previous rows at the top of the loom (beaten upward), using the fingers, a bone weaving comb, or an iron "weaving sword" with a long, flat blade.

In a more advanced version of the warp-weighted loom, the process was simplified by the introduction of the "heddle," a device that made it possible to raise a complete set of warp threads with a single movement. The odd and even warp threads hung down alternately in front and in back of a fixed horizontal "shed rod." A second, adjustable bar called the heddle was loosely joined by loops of twine to the rear warp threads. The weaver first passed the weft through the "natural shed," the space between rear and front warp threads created by the shed rod. On the next row, she moved the heddle forward in its brackets, pulling the rear warp threads to the front and creating a space known as the "artificial shed," through which

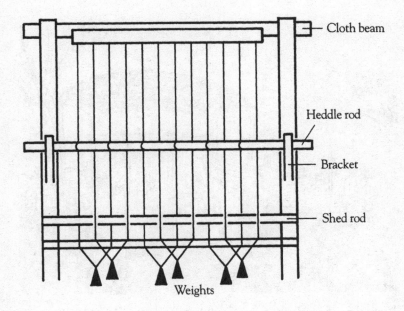

Cloth beam

Heddle rod

Bracket

Shed rod

Weights

Vertical warp-weighted loom (schematic drawing).

the weft was passed. Alternating these two positions and sheds produced plain weave. Variations in pattern could be created by changing the arrangement of warp threads and by increasing the number of heddle rods.[43]

The vertical two-beam loom was operated similarly, except that it was usually smaller and narrower, and the weft was beaten downward instead of up. The Utrecht Psalter (c. 834) shows a woman in a gynaeceum weaving outdoors under a canopy, using a two-beam loom, separating the warp threads with her fingers.

Linen was spun and woven by the same processes as wool, but the raw fibers required a more extensive treatment: first hanging the bundles of flax to dry so that the seeds could be shaken out, then "retting"—soaking in water—and pounding to remove the bark, finally "hackling," drawing the stalks across a board set with rows of spikes to remove the rest of the stem and to separate the fibers.[44]

Vertical two-beam loom, from a twelfth-century copy of the Utrecht Psalter. [Trinity College, Cambridge, Ms. R 17, 1, f. 263.]

"Men of Iron"

In contrast to cloth making, military technology was an area that in the early medieval centuries experienced radical transformation, embracing weapons, defensive armor, fortifications, and, historically most intriguing, the equipage of a riding horse, especially the stirrup.

At the outset of his career, the late Lynn White proposed a bold hypothesis: the stirrup, imported from Asia, made possible shock combat by mounted knights, whose endowment by Charles Martel with Church lands to pay for their expensive gear laid the foundation for feudalism.[45] White's essay stirred controversy, research, and critical analysis that yielded a more complex picture. It is now established that the campaigns of Charles Martel and Charlemagne were dominated by sieges and raids, with little evidence of shock combat, and that a nobility of birth already existed, with origins in the old Frankish aristocracy; it merely gained an infusion of blood from the knights, who appeared on the scene in the tenth century. The foundations of feudalism included customs of both Germanic and Roman society, and the system reached maturity only in the thirteenth century.[46]

Nevertheless, if it was not the catalyst White suggested, the stirrup had military impact and social repercussions sufficient to justify the term "revolution." Its beginnings trace to India in the second century B.C., in the form of a loop into which the rider thrust his big toe. Such a stirrup could give only slight assistance to staying on the horse and even less in mounting, besides being limited to barefoot, warm-weather riders.[47] Iconographic evidence of the true stirrup dates from the early fourth century A.D. in China, whence, like so many innovations, it gravitated westward. Turkish Avars, who appreciated its steadying effect in firing arrows from the saddle, brought it to Hungary, whence it passed to the rest of Europe, evidently valued mainly for assistance in mounting: the words for stirrup in Old High German, Old Saxon, and Old English all derived from words for climbing (heretofore horsemen had used a mounting

*This ninth-century equestrian statue of
Charlemagne shows no stirrups. [Louvre.]*

stool or vaulted onto horseback). The earliest representation of
a stirrup in the West occurs in a St. Gall manuscript of the late
ninth century; the Utrecht Psalter of circa 834 shows many
mounted warriors but none with stirrups.[48]

When European horsemen finally adopted the stirrup and
matched it with the contoured saddle, they gained a dramatic
advantage. From their newly secure seat, they could deal heavy
blows, at first with the existing battle-ax, later with long sword
and heavy lance. The last weapon especially created true "shock
combat" by permitting a blow to be struck with the energy
derived from the mass of the charging horse. How much the
advantage was used, and when (outside tournaments) is still an
open question. In the Bayeux Tapestry of circa 1080, mounted
combatants on both Norman and English sides are shown hurling spears and lances, rather than driving them couched.[49]

Besides the nailed horseshoe, which arrived from the East at

Norman knights at the Battle of Hastings (1066) are equipped with stirrups, but throw their spears. Bayeux Tapestry. [Phaidon Press.]

about the same time as the stirrup, Europeans added to cavalry accoutrements two native inventions, spurs and the curb bit, providing effective control for a rider who had only one hand free for the reins.[50]

Fighting from horseback encouraged the adoption of heavy defensive armor, which quickly gave cavalry the ascendancy on the battlefield. Charlemagne's biographer Notker describes the formidable appearance of Charlemagne and his army at the siege of Pavia (774), in full battle gear:

> That man of iron [was] topped with his iron helmet, his fists in iron gloves, his iron chest and his broad shoulders clad in an iron cuirass. An iron spear raised on high against the sky was gripped in his left hand. In his right he held his still unconquered sword . . . [His thighs] were bound in plates of iron . . . his greaves [lower-leg coverings] too were made of iron. His shield was all iron. His horse itself gleamed iron in color and in mettle. All those who rode before him, those who accompanied him on either flank, those who followed, wore the same armor, and their gear was as close a copy of his own as it is possible to imagine . . . The rays of the sun were reflected by this battle-line of iron. This race of men harder than iron did homage to the very hardness of iron.[51]

In picturing a whole army clad in plate armor, Notker exaggerated; plate armor was for kings and leaders. The universal armor of the ordinary mounted soldier of the early Middle Ages

*Shock combat: knights charging with
couched lances, from the twelfth-century
Life, Passion, and Miracles of St.
Edmund. [The Pierpont Morgan
Library, M. 736, f. 7v.]*

was "mail"—metal scales, strips, or rings sewn on a leather or
padded-cloth tunic. A coat of mail was expensive; so were the
helmet, shield, and arms, not to mention the large, specially
bred horses (chargers, destriers); in the marketplace such a
horse was worth as much as four to ten oxen or forty to a hundred sheep. A coat of mail, made up of tens of thousands of
individually forged iron rings, was worth sixty sheep.[52] As the
cost of equipment rose, a social transformation followed the
military one, and from soldier of mediocre status the knight was
elevated to member of a prestigious caste, graced with a code of
conduct that exerted strong influence on posterity.

The military landscape of the ninth century featured another
technological innovation that became a symbol of the Middle

Ages: the castle. The disorders that followed the disintegration of Charlemagne's empire, exacerbated by the Viking raiders, made northwest Europe look to its defenses. Towns rebuilt long-neglected walls, reviving the half-forgotten art of stonemasonry, while in the countryside, fortresses appeared, but fortresses of a novel description. Masonry was too costly for the thinly populated rural districts, and the new structures were of timber and earth—"motte and bailey," the motte a mound, natural or artificial, the bailey a palisaded court below. Even more distinctive than their physical form was their social character. Public forts manned by professional garrisons were of long standing, but the motte-and-bailey castle was a private fortress, not exactly by design but by an inevitable progression. Originally intended as command post for a Carolingian imperial officer, the "castellan," who lived in it with his family, servants, and retainers, it soon became an independent hereditary possession and the castellan the ruling authority of his local district.

Cheap and quick to build, requiring little skilled labor, the motte-and-bailey castle was nevertheless militarily effective. It could not only block the invasion of a region but control the local population. Chronicler Jean de Colmieu describes the building of a motte:

> It is the custom of the nobles of the neighborhood to make a mound of earth as high as they can and then encircle it with a ditch as wide and deep as possible. They enclose the space on top of the mound with a palisade of very strong hewn logs firmly fixed together, strengthened at intervals by as many towers as they have means for. Within the enclosure is a house, a central citadel or keep which commands the whole circuit of the defense. The entrance to the fortress is across a bridge . . . supported on pairs of posts . . . crossing the ditch and reaching the upper level of the mound at the level of the entrance gate [to the enclosure].[53]

In times of peace, the lord and his family lived in the keep on top of the mound, the garrison, horses, and other livestock in wooden structures in the bailey (courtyard) below. Threatened with attack, everyone withdrew to the keep.

Remains of motte-and-bailey castle built by William the Conqueror at Berkhamsted. The motte (mound) was topped with a timber stockade. Berkhamsted was unusual in having a wet moat. [Aerofilms Ltd.]

The economy and effectiveness of the motte-and-bailey castle led to its survival into the high Middle Ages, in competition with its more elaborate and far costlier masonry counterpart. A thirteenth-century motte-and-bailey described by the priest Lambert of Ardres had three stories: on the ground floor, storage rooms; on the second, the hall where the family lived and ate, the "great chamber" where lord and lady slept, and the nursery; on the third, a dormitory for the adolescent children and the servants, and a chapel.[54]

At the other end of Europe, where the surviving eastern half of the Roman Empire, now frankly Greek, or Byzantine, was beset by enemies, a piece of military and naval technology more dramatic than the stirrup suddenly appeared in the seventh century. If "Greek fire" alone did not preserve Byzantium, it certainly helped.

Incendiary weapons were not themselves new to warfare. Naphtha (a petroleum distillate) was known as early as the fourth century B.C., and petroleum, sulfur, bitumen, and resin were used in both land and naval warfare in the first centuries of the Christian era. The new Greek mixture, credited by the Byzantine historian Theophanes to Callinicus, a Syrian refugee from the Arab conquest, was discharged from tubes mounted in ships' prows and could not be extinguished with water. Its chemical composition has defied positive analysis, and the method of ignition is even more puzzling. Probably it was a mixture of distilled petroleum, not unlike modern gasoline, thickened with resinous substances and sulfur to slow its dissipation in water and keep it from being washed away from the target by wave action. Reports that it ignited on contact with water were probably the result of faulty observation. A similar Chinese weapon of the tenth century, "fierce fire oil" (distilled petroleum), was ignited by a charge of gunpowder, but the Byzantines had no gunpowder in the seventh century. It has been conjectured that they projected their mixture with a force pump, the Hellenistic device credited to Ctesibius. Still another puzzle is how the Byzantine sailors managed to store so volatile a mixture safely aboard their own ships. A modern conjecture is that the mixture was not volatile until heated and pressurized below decks immediately before combat.

The secret was the more easily guarded because of its inherent complexity, comprising not only the mixture but method of preparation and means of discharge. The weapon was first used, with devastating effect, against an Arab fleet attacking Constantinople in 673, and again in 717. An eighth-century account describes iron shields that protected the men who worked the bronze flamethrowers, and the thunderous noise of the flaming jets, sometimes hand-held, sometimes mounted on the ships. "Thanks to the cooperation of God through His wholly immaculate Mother's intercession," recorded the chronicler Theophanes, "the enemy was sunk on the spot. Seizing booty and the Arab's supplies, our men returned with joy and victory." Eventually the Arabs acquired at least a version of the weapon, while its secret, limited to a very small ruling circle, was meantime (before 1204) lost by the Greeks themselves.[55]

Swords and Plowshares

Notker's description of Charlemagne as "that man of iron" reflects the aristocratic role the "democratic metal" had assumed in early medieval Europe. Metal mining and production of all types had fallen off sharply with the loss of the powerful sponsorship of the Imperial government, and some of the old Greco-Roman techniques had been lost. But small-scale open-pit mining and local forges continued operations on both sides of the old Roman frontier, and as the population grew, the need for more and better agricultural equipment, such as the new heavy plow, supplied a stimulus in addition to that of the knightly "men of iron" with their swords, battle-axes, helmets, and chain mail.

The medieval blacksmith enjoyed high prestige, including the reputation of being in league with diabolic powers; a smith might be consulted to cast or break spells, to cure disease, or to repair broken bones. As a fabricator of armor and weapons, in Wales he sat at the table next to the royal chaplain, and at Tara, the ancient Irish royal hall, he joined the royal verse maker, brewer, and teacher on the king's right hand.[56] In time the allied professions of armorer and locksmith became spin-off specialties, and the blacksmith limited himself to farm-implement manufacture and repair, nail making, and horseshoeing. Throughout the early Middle Ages he remained the most important of the skilled craftsmen. The tenth-century Anglo-Saxon writer Aelfric composed an imaginary debate among craftsmen on their relative importance in which the smith triumphantly pointed out that none of the others could work without the tools he made them.[57]

Hardly any relevant documents survive on early medieval metallurgy, and though archaeological sites have turned up a wealth of evidence, such as the twenty-four-furnace array at Zelochovice, Czechoslovakia, dating and interpretation are difficult. The first step in smelting was washing and roasting the ore and breaking it into chunks suitable for the reduction furnace. After centuries as a mere depression in the ground with clay lining and dome, the furnace began its rise by adding a stubby chimney of clay and sandstone. Besides the gas exit, two openings were provided, a charging hole for introduction of the ore and an aperture near the bottom to allow extraction of the "bloom" of iron—soft and glowing but not molten—and introduction of a draft supplied by a pair of bellows. In its appetite for fuel, the furnace was rapacious, typically burning twelve pounds of charcoal to smelt a pound of iron.[58]

The hot bloom was pummeled on a flat stone to expel the slag (sand and clay) and consolidate the iron into bars and plates suitable for the smith. The master ironworkers "only knew how, they did not know why, they did certain things" (W. K. V. Gale). They knew that their charcoal created iron; they did not know that it did its job by drawing the oxygen out of the ore (iron oxide) to unite with carbon and escape as a gas. They knew that if the heat was maintained too long, something went wrong; they did not know that the iron having got rid of its oxygen began taking on carbon. Much later, when furnaces produced high enough temperatures to melt iron, the molten metal could be cast in molds (hence the name "cast iron"), but in the early Middle Ages the accidental production of solid lumps of metal too brittle to be worked was merely a nuisance. If all went well, several hours' work produced a few pounds of usable iron.[59] Certain ores, rich in manganese, produced a naturally steely iron.[60]

The medieval bellows consisted of a wooden box closed by a piece of leather that could be pressed by hand or foot, forcing the air out of a small hole into a pipe leading to the tuyere (nozzle) and thence to the fire. A cord and a springy piece of wood pulled the bellows skin up again, admitting a fresh charge of air. In the smelting furnace, bellows were paired, so that their alternate blows kept a steady draft going.

The smith who received the iron from the reduction furnace relied, like the iron smelter, on his skill, experience, and the knowledge passed from generation to generation at the forge over a thousand years. He knew how to harden his metal by reheating it to a high temperature and holding it there for a certain length of time, not why the effect occurred: the iron was absorbing from the charcoal a small amount of carbon (less than one percent), and here too if the process was allowed to go on too long, the product became brittle and unusable. He did not even realize that the hardening he achieved was only surface deep—"case hardening" in modern metallurgical vocabulary—only that the improvement made the metal far more valuable.[61]

More and more iron now went into plowshares, harrows, sickles, and billhooks, and even some church bells were made of iron, the smith welding or riveting four pieces together. But the most famous products of the smith were the formidable long swords wielded by the armored knights, produced by stretching and refolding the case-hardened strips back on each other. The finished blade had a characteristic appearance, "rather like streaky bacon" (Leslie Aitchison), and a high price.[62] In Charlemagne's time a good sword cost as much as three cows and was correspondingly treasured. Its owner might give it a name and even have religious relics enclosed in the hilt. A legendary character came to be attached to the swords of great heroes: Roland's Durandal, Charlemagne's Joyeuse, King Arthur's Excalibur, El Cid's Tizona.[63]

Chain mail, known in ancient times, may have had an Eastern origin, but medieval smiths and armorers fabricated it in previously unheard-of quantities. How it was made is not certain; the mature version of the hauberk or coat of mail consisted of hand-drawn iron wire soldered into rings, each ring linked to its neighbors. The rings-sewn-on-leather version was retained for foot soldiers and archers.

More and better iron improved the smith's own tools and those of his frequent partner, the carpenter. Few new devices were introduced, but many old ones became for the first time widely diffused, including one machine very important for woodworking and carpentry, the lathe. Designed to rotate a workpiece rapidly against a cutting tool, the medieval version

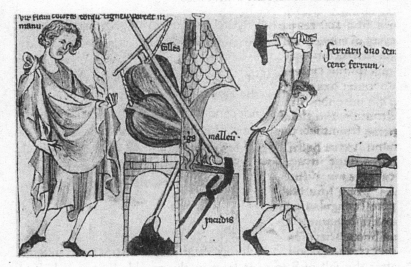

Waist-level hearth with bellows. [British Library, Sloane Ms. 3983, f. 5.]

came in two forms, the pole lathe and the bow lathe. In the first, a cord was wound around an axle and fastened by one end to a pole, which bent downward as the other end of the cord was pulled to rotate the axle; released, the pole supplied spring action to spin the axle in the opposite direction. The bow lathe substituted a bow, fixed overhead, for the pole.[64]

The massive swords and battle-axes of the men of iron required sharpening when first fabricated and at intervals afterward. So did many of the agricultural and household implements. For long ages, blades had been honed by reciprocal rubbing against a naturally abrasive stone (usually sandstone). The Chinese had invented a rotary grindstone much earlier, but whether the idea migrated westward or originated independently in Europe is unknown. Its significance lies in its character as the earliest application of the crank principle, and its first representation in Europe is in the Utrecht Psalter (834), illustrating the Sixty-third Psalm: a rotary grindstone is being operated by a man turning an unmistakable right-angled crank. Lynn White pointed out that the wicked ("those that seek my soul") are depicted sharpening their swords on an old-fashioned whetstone, while the righteous, preparing to battle them, sharpen theirs with the new, technically innovative rotary grindstone.[65]

One of the simplest and historically most valuable of machine components, the crank presents a mystery in its slow diffusion westward or independent invention and reinvention. From the rotary grindstone, the medieval crank moved to the hand organ or hurdy-gurdy (originally a church instrument), then to the crossbow and the brace and bit, before at last, nearly at the end of the Middle Ages, becoming a machine component.[66]

The Romanesque Church

Like mining and metallurgy, building construction suffered after the fall of Rome, at least in the troubled western half of the Empire. In the eastern half, sixth-century Constantinople acquired an elaborate sewage system and the magnificent domed cathedral of Santa Sophia, which rose in the incredibly short time of five years.[67]

In the early medieval West, lords and prelates stripped the Roman temples and public buildings of their marble facings and despoiled walls of their bricks to ornament their own palaces,

Illustrating the Sixty-third Psalm, a rotary grindstone is operated, lower right, while man at lower left sharpens sword on old-fashioned whetstone. From the ninth-century Utrecht Psalter. [Utrecht University Library, Aev. med. script. eccl. 484, f. 35v.]

churches, and country houses. Stonecutters turned to odd jobs and ceased transmitting their skills, while brick making disappeared from northern Europe until the high Middle Ages. Charlemagne, restoring some of the majesty of Imperial power, had difficulty in rounding up enough skilled masons to build the royal palace and chapel in his new capital of Aachen. Masonry structures of the Carolingian Age show a reliance on mortar, troweled thickly on the stone courses, in contrast to the elegant Roman practice of using no mortar at all, or thin layers only. The secret of Roman cement and concrete was lost, not to be recovered until the nineteenth century.[68]

Financial and technical constrictions notwithstanding, the early Middle Ages did a considerable amount of building. Greek and Roman temples had been rarities, the Roman religion being practiced mainly at home rather than in a church. The triumph of Christianity, institutionalized in community worship, demanded churches in numbers the Western world had never seen. The "basilicas" designed to accommodate city congregations were large, built of stone, and modeled on the Roman public hall, divided by rows of columns into nave and aisles. In the sixth century, Gregory of Tours recorded the dimensions of the new basilica of Tours, built over the tomb of St. Martin, as 160 feet long by 60 feet wide and 45 feet high, with 52 windows, 8 doors, and 120 marble columns. The new cathedral in Gregory's native Clermont was equally impressive.[69]

The monastic movement launched by St. Benedict also required large buildings, such as the abbey of Jumièges in Normandy (seventh century) and the abbey church of St. Denis, near Paris, rebuilt from a fifth-century church in about 760. Nearly all the "Romanesque" (Roman-like) churches were built on the plan of the basilica, often with a transept projecting on either side and a rounded apse at the eastern end. Churches in Gaul often had bell towers, in the middle, over the crossing, or as separate structures. Charlemagne's renaissance stimulated the rebuilding of many Merovingian churches on a still larger scale, for example, at Cologne and Rheims.[70]

Most churches were merely roofed with timber, but some

were provided with vaulting, in the form of the tunnel-like barrel vault, which with its massive supporting walls cost fifteen to eighteen times what timber roofing did and permitted only small window openings. Yet in the meager windows of the early medieval basilica the first stained glass appeared. The historian Bede describes the glazing of the windows of monastic buildings at Monkwearmouth (County Tyne and Wear) in the seventh century by craftsmen imported from the Continent, to ornament "a stone church built ... in the Roman style."[71] By the tenth century, stained glass (actually not stained but emerging colored from the glassmaking process) was already a glory of Christian cathedrals.

Rivers, Roads, and Bridges

In addition to the Christian churches, a few Roman-type public works were built in the ninth century. Charlemagne restored

Interior of ninth-century abbey church of St. Philibert, near Nantes, showing heavy piers of the nave.

the famous Roman lighthouse at Boulogne, and St. Aldric, bishop of Le Mans, caused a 4½-mile-long aqueduct to be constructed to serve his city, while along the upper Loire, levees were erected to aid navigation.[72]

The rivers of northwest Europe, comparatively free from winter ice and flowing year round, served commerce well even in a natural state. They carried a substantial part of the increasing traffic in wine, which benefited from the now general substitution of a variety of wooden barrels, vats, and tubs for the amphorae and skins of the ancient world. Like the Mississippi flatboatmen of a later era, medieval river men poled their craft downstream, sold them for timber at their destination, and walked back. Where water traffic was heavy, oxen trod towpaths hauling barges.[73] Charlemagne conceived a grandiose project for a canal linking the Danube with the Rhine, which was actually started but left unfinished until a thousand years later.[74] Whether ninth-century engineers could have solved the problem posed by the different levels of the two rivers is conjectural; the canal lock did not yet exist, even in China.

Lords who enjoyed locations on busy rivers copied the Roman authorities in levying tolls but not in applying the revenue to improve navigation. Excessive tolls actually reduced traffic in some cases, and some lords were no better than pirates, or, in the phrase they added to Western languages, "robber barons."

Roads were less important in commerce than rivers. Hardly anything is known of early medieval road building and maintenance except by implication. Roman roads probably deteriorated more quickly in the wet, cold northern climate than in the South, but all roads require extensive maintenance, for which no adequate governmental or even regional authority now existed. Barbarian kings who inherited the Roman Empire's taxing power often allowed it to lapse out of indifference, their traditional governmental responsibilities imposing no demand for large revenues. The need for road maintenance only slowly gained the stature of a government problem.[75]

A further complication arose as the northward and westward

movement of agriculture and settlement created whole new route orientations, notably from the old Lyons-centered network of Roman Gaul to the Paris-centered web of Merovingian France. Unused pavements acquired overlays of soil and vegetation, under which they slumbered until disturbed by modern archaeology. The Roman road that best retained its importance was that traveled by English pilgrims on their way to Rome, crossing France from Boulogne to Langres to the Alpine passes, a route preferable to the risky sea voyage even in the ninth century, when the passes were infested with Saracen bands.[76]

"Incomparably more important" than roads in the improvement of medieval transportation and communication, according to Marjorie Boyer, were bridges.[77] Of the three kinds of river crossings in existence, fords were often unfordable, ferries inadequate, bridges, if they existed, in need of repair. The general substitution of permanent, well-maintained bridges for fords and ferries has been called (by C. T. Flower) "the great public work" of the Middle Ages.[78] Pack animals could negotiate even a bad road, but not a river served only by a skiff, or nothing at all. In the obscurity of the early Middle Ages, at least a beginning was made in solving the river-crossing problem.

All that we know of early medieval spans is that they were made of wood and stone, certainly more wood than stone. Descriptions are entirely lacking, and hardly a fragment survives. We do know that Charlemagne and his successors imposed responsibility for bridge building and maintenance on the local inhabitants, with a military function primarily in mind. A fortified bridge could not only block a crossing but control traffic in the river. A bridge over the Marne at Treix prevented the Vikings from ascending the river. The problem of financing construction was met by a double collection of tolls, for both land and water traffic.

As early as the sixth century, supplemental roles were found for bridges. One important one was conceived during the siege of Rome by the Goths in 537, when the enemy shut off the aqueducts whose water drove the city's gristmills. Belisarius, the

Byzantine general defending the city, ordered floating mills installed close to the Tiber bridges, whose piers constricted and accelerated the current. Two rows of boats were anchored with waterwheels suspended between them. The arrangement worked so well that cities all over Europe were soon copying it.[79] The Grand Pont in Paris, probably a combination of wood and stone, built water mills under its roadway and houses on top of it, inaugurating one of the Middle Ages' most picturesque architectural fashions.

Navigation: Lateen Sail and Long Ship

While land transportation struggled with roads and river crossings, at sea two new shipbuilding traditions emerged, the northern and the southern. By the year 900, both had achieved important advances in design over the ships of the ancient world.

In the Mediterranean the triangular (or near-triangular) lateen sail was at last successfully rigged on large vessels. The Byzantines, who may have learned the technique from their Muslim enemies, passed it on to other Europeans, and by 800 it had become the dominant sail on the Mediterranean. Hung from a long, sloping yard, one end of which rose well above the masthead while the other reached nearly to the deck, the lateen could take the wind on either side, improving a ship's ability to sail close to the wind. This was especially valuable in helping a ship fight off a lee shore. The old picture of medieval sailing ships hugging the coast is a misconception, since the coastal rocks and shoals represented the main danger to a vessel. But much navigation in coastal waters was unavoidable.[80]

Manipulating the lateen was not easy. To come about (change tack) required lifting the yard over the top of the mast, calling for extra skill and extra manpower, but the advantages outweighed the difficulties. The Byzantine navy's sail-and-oar "dromons" made good use of the lateen in getting into position to project their Greek fire.[81] Though designed specifically for war, the dromon also proved a practical cargo carrier.

Simultaneously with adoption of the lateen sail, Byzantine

and other Mediterranean shipbuilders introduced a radical new system of hull construction. After long following the Roman method of building up the hull plank by plank, each succeeding plank fastened edge to edge to its predecessor by mortise and tenon, and the supporting skeleton inserted late in the job, they now reversed the procedure and built the skeleton first. Much skilled labor was saved, and though the resulting hull was not as strong as the Roman, the savings in labor translated into cost savings for carrying commercial cargo.[82]

Meanwhile, northern European shipbuilding was pursuing its own long line of evolution, entirely isolated from the Mediterranean tradition. Two very early specimens of northern ships, one prehistoric, the other from the Roman period, have been discovered by archaeologists. The older, found at Als, Denmark, dating to about 350 B.C., was forty feet long and clinker built, that is, with its planks overlapping, lashed together rather than nailed. The ship had no mast or keel and no oarlocks, and was evidently paddled by its crew, canoe fashion.[83] The second, recovered from a burial site at Nydam, Schleswig, and dating from the third century A.D., was seventy-five feet long, its clinker-built planks nailed as well as lashed, making the hull both flexible and durable. Still lacking a mast and limited to coastal navigation, the Nydam ship embodied a large advance in power: its twenty-eight crewmen faced sternward and pulled on oars.

These premedieval examples foreshadowed the configuration of an eighty-foot-long Anglo-Saxon ship found at Sutton Hoo, England, near Ipswich, dating from about A.D. 600, clinker built but by a more advanced technique, its short planks riveted together lengthwise to form long strakes (ship's-length timbers), eliminating the need for hard-to-find long, straight tree trunks. Still lacking mast and keel, the ship was apparently used for coastal and cross-Channel navigation, probably in the trade partnerships with the Merovingian kings. The space taken up by the two dozen or more rowers, however, left little room for cargo.[84]

By this time, northern ships probably included sailing craft, but no evidence of them has survived. The first northern ship to combine sail and oar propulsion that archaeology has uncovered is a late-ninth-century Viking vessel found at Gokstad, Norway. Its keel was fashioned from a single giant oak trunk, providing great strength and supporting a mast anchored firmly in a strong socket (keelson). Such a ship could sail all night in the open sea before a favorable wind, vastly increasing its range and potential.[85]

The principal weakness of the galley had always been its lack of sleeping room for the crew, virtually imposing nightly landings. A ship equipped only with oars had to inch its way along the coast to the English Channel and home again to Denmark or Norway. An open-sea sail-and-oar ship could reach the Channel coasts from Scandinavia in a few days and there exploit its shallow draft (about 3½ feet) and maneuverability by ascending rivers to trade or raid.

The deck of a Gokstad-type ship was composed of loose planking, which kept out the sea but was easily opened to load cargo. Its overlapping hull timbers were secured by wooden pegs (treenails) driven through holes bored to receive them, the pegs split and wedged for snug fit; gaps between the planks were made watertight by pressing in moss. Its oar holes were keyhole shaped, to permit oars to be passed out from inside the ship, and equipped with shutters to close them against the sea when oars were shipped. Like its predecessors, it could readily be beached for landing, loading, and unloading, taking advan-

Reconstruction of the Anglo-Saxon Sutton Hoo ship, c. A.D. 600. [British Museum.]

tage of the tides. The mast, probably at first immovable, could by the ninth century be unstepped and the square sail shifted to face in different directions, even fore and aft. Clumsier than the lateen sail, it nevertheless served its purpose.[86] A nine-teenth-century replica of the Gokstad ship sailed from Bergen, Norway, to Newfoundland in twenty-eight days.[87]

The product of a lengthy evolution, the longship of the Gokstad type was not deliberately designed for raiding but turned out to be ideal for the purpose. Its maturing coincided with a moment when Europe was both a vulnerable and an inviting target. Though ethnically and culturally homogeneous, the Vikings were geographically divided in three, a circum-stance that led them in three different directions. Those from Norway tended to sail west to the Shetlands, then turn south into the Irish Sea, thence to the French and Spanish coasts. Those starting from Denmark turned into the English Channel, raiding Britain and the Low Countries, while those from Swe-den crossed the Baltic to Germany, Lithuania, and Russia. The hit-and-run character of Viking (as well as Saracen) raids foiled the uncoordinated defensive efforts of Europe's dispersed politi-

Model of the Gokstad ship, c. A.D. 900. [Science Museum, London.]

cal and military authority. Loaded with plunder, the longship turned merchant-trader, often setting up temporary shop in a district next door to the one it had just victimized. In the Viking mixture of raiding and trading, trading gradually came to predominate, mainly in the form of an extensive commercial empire in parts of eastern Europe and in Russia. Here the Vikings met Muslim merchants coming up the rivers from the other direction and traded furs and slaves, weapons and amber for the Muslims' silks, spices, and silver coins.

A Muslim diplomat described the arrival of Viking ships at a trading post on the Volga: the Northerners were "tall as date-palms, blond and ruddy," never parted from "axe, sword and knife . . . the filthiest of God's creatures . . . lousy as donkeys." Each trader went ashore to make an offering before "a large wooden post with a face like that of a human being," placing bread, meat, onions, and beer at the foot of the image and then announcing, "Oh, my Lord, I have come from a far land and have with me such-and-such a number of slave girls and such-and-such a number of sables," and enumerating his other wares, then praying the idol to "send me a merchant with many dinars and dirhems, who will buy from me whatever I wish and will not dispute anything I say."[88] Returning to northern Europe, the Vikings exchanged the silver dinars and dirhems for Rhenish wine, glassware, querns, and weapons, with the Frisians acting as go-betweens.

Meanwhile, in the West, resistance finally hardened, and the Scandinavian rovers gave up piracy and settled down in England and Normandy to crop and stock farming, supplemented by commercial dealings. The famous longship, however, was less well adapted to peaceful pursuits than to raiding and plundering. Better fitted to carry cargo—or settlers—especially on long voyages, was the Viking version of the round ship, the "knorr." Broad in beam, with six or eight oars positioned at the ends of the ship but with main reliance on sail, the knorr required a much smaller crew than the longship.[89] It was clinker built, but by the tenth century clinker construction was obsolescent. A better round ship had already appeared, originating in

Sluys, in the Low Countries. The "cog" employed carvel construction, borrowed from the Mediterranean, and a flat-bottomed hull design that helped in navigating the shoals of the southern shore of the North Sea and facilitated beach landing on sandbanks.[90] Both cog and knorr found a valuable new role in the mysterious migration into the Baltic of herring, unknown there in ancient times. Preserved in salt, herring made an excellent cargo for the English Channel and Bay of Biscay, where salt could be taken on as cargo for the return voyage.

Northern navigation, like southern, continued to suffer from serious deficiencies: inadequate rigging, weakness of hulls against storms, lack of maps, charts, and instruments. Both northern and southern sailors stayed home through the winter, except in rare cases. One such was recorded by St. Willibald, who was hardy enough to sail from Tyre, in Syria, to Constantinople, a coastwise distance of perhaps a thousand miles, in November of 726. He arrived the following spring, a week before Easter.[91]

"Nature's Secret Causes"

Commenting on the intellectual climate of the early Middle Ages, science historian R. J. Forbes has written, "Science was no longer the handmaid of philosophy; instead both were made to serve religion."[92] Philosophy in the early Middle Ages was in fact virtually indistinguishable from religion, and science was indeed its handmaid. Medieval theologians would have been surprised to learn that there was any difference among the three spheres of thought. Although St. Augustine believed that man could best perceive God directly through faith rather than through observing the created world, most medieval thinkers were followers of an alternate tradition, that of Boethius, who believed that knowledge of God could be attained through examination of the beauty and order of the universe. The man of reason, Boethius wrote, "sought the causes of things—why the sighing winds vexed the sea waves, what spirit turns the stable world, and why the sun rises out of the red East to fall beneath the western ocean . . . what tempers the gentle hours of spring . . . what causes fertile autumn to flow with bursting grapes in a good year," in short, "nature's secret causes."[93]

Boethius conducted experiments with the monochord, the single-stringed instrument that had been used by Pythagoras to demonstrate the mathematical order of the universe. Dividing the string proportionally produced the notes of the Pythagorean scale, which the Greeks believed represented divine proportions existing elsewhere in the universe, for example, in the distances separating the planets. Boethius and his successors elaborated on the Pythagorean conception of a mathematical universe, with the conviction that it must have (in the words of John Benton) "a musical, harmonious, beautiful, and ordered pattern . . . if only one can find it."[94]

In the service of science-philosophy-religion, medieval churchmen did much to conserve knowledge (at least Latin knowledge) and even to add to it. Theologians such as Isidore of Seville, Hrabanus Maurus, Martianus Capella, and John Scotus Erigena devoted themselves to expositions of learning, their works providing students for the next several centuries with textbook information (not all of it reliable) on the liberal arts, medicine, agriculture, natural history, astronomy, and other secular topics. The Venerable Bede (673–735) described the earth as a sphere, which like his ancient predecessors he located in the center of the universe, and gave a lucid account of eclipses and phases of the moon. He also composed a history of the world, for which he invented a new system of dating: backward and forward from the birth of Christ.

Charlemagne's counselor Alcuin (732–804) promoted the adoption of "Caroline [Carolingian] minuscule," a more readable and regular version of several lower-case scripts invented by the clergy in the preceding century to replace Roman upper case (majuscule) in order to economize on expensive parchment or vellum. The new script facilitated the copying of the Latin classics, rescued from oblivion by the monasteries, "islands in a sea of ignorance" (Charles Homer Haskins).[95] Caroline minuscule came to be universally adopted in the West,

Boethius, from a manuscript of On the Consolation of Philosophy. [Bodleian Library, Ms. Auct. F 6.5, f. 1v.]

where a version of it eventually contributed an essential element to the invention of printing: the typeface. Most typefaces in use today are descendants of Caroline minuscule.

The monasteries manufactured their own writing materials, trimming the skin of a calf, sheep, or goat, soaking it in lime, stretching, scraping, and cutting it into sheets. A better ink introduced in the seventh century, gall-iron ink, made with a soluble iron salt, supplanted the old carbon-black-and-gum solution. The scribe, sitting on a bench with his feet on a low stool, worked at a writing desk equipped with inkhorn, quill pen, and erasing knife, copying sometimes directly from a manuscript, sometimes from the dictation of a reader. To test their pens, scribes jotted comments in the margins: "How hairy this parchment is!" "It is cold today." "It's dinnertime."

The scribe Eadwine, from the twelfth-century Canterbury Psalter. [Trinity College, Cambridge, Ms. R 17, f. 1.]

Manuscripts commissioned by the wealthy required the additional services of a painter, who decorated the initials and added other embellishments in colored inks and gold leaf.[96]

The Christian clergy also served science and technology as teachers. The Benedictine monk was "the first intellectual to get dirt under his fingernails" (Lynn White),[97] as he explored and taught agricultural science, stock breeding, forestry, metalworking, glassmaking, and other useful arts. The great monasteries maintained their corps of artisans. The abbey of Corbie, near Amiens, had six blacksmiths, a fuller, two goldsmiths, a parchment maker, and four carpenters; St.-Riquier, on the Channel coast northwest of Corbie, also had shoemakers and saddlers. The famous plan of St. Gall (Switzerland) shows workshops for cabinetmakers ("turners"), harness makers, saddlers, shoemakers, blacksmiths, goldsmiths, fullers, and sword polishers.[98]

EUROPE 900

The new motte-and-bailey castles dominating local regions visually dramatized the developing political order, known to future historians as feudalism. As Vikings and Saracens retreated, the local lords could turn their attention and their innovative military technology against each other, creating a sort of Europe-wide anarchy on horseback. "The strong built castles, the weak became their bondsmen" (James Bryce).[99]

But under the untidy surface, more meaningful change had taken place. By the beginning of the tenth century, notwithstanding the fall of Rome, Vikings, Saracens, and the loss of Greek science, the new Europe had in its technology clearly surpassed the ancient Mediterranean world. In agriculture, metallurgy, and sources of power it had introduced significant improvements, inherited, borrowed from Asia, or invented independently. Its continuing demographic surge was beginning to be reflected qualitatively, in the growth of cities. Slavery, decaying as an institution, no longer supplied the basis of agricultural labor, and the Church pronounced emancipation of slaves to be "good work par excellence."[100] (On the other hand, the Christian religion lent a moral cover to the slave trade, limiting it to non-Christian Slavs and Arabs, while Islam limited it to non-Muslims.) For everyone the standard of living was rising—in Robert Reynolds's words, "not from high to higher, but from very low to less low."[101]

The revolution in agriculture that introduced new implements, new techniques, and a new organization of work was largely a revolution from below, not above. "The hero of the late Roman Empire and the early Middle Ages," wrote Lynn White, "is the peasant, although this cannot be discovered from Gibbon."[102]

Less conspicuous than the castle but more significant for the long future was the above-ground reduction furnace, feeding iron to local forges whose smiths shaped it into parts for plows, spades, pitchforks, and shoes for horses beginning to pull with the aid of the new horse collar. As the horse trod Europe's fields to cultivate the crops, the waterwheel turned "at wondrous speed" to grind the grain, while the triangular lateen sail drove ships on the Mediterranean—three more symbols of progress in a not so Dark Age.

4

THE ASIAN CONNECTION

THE GIFTS TO EUROPE IN THE EARLY MIDDLE Ages of silk culture and production, padded horse collar and stirrup, and perhaps the crank were part of a long history of technology transfer among regions and continents dating back to Neolithic times. Such long-distance exchanges are indicated by similar weapons, tools, and religious practices found by archaeology among Eskimos (Inuit), North American Indians, Chinese, and Siberian peoples. Later the great Bronze Age inventions of Egypt, Mesopotamia, and the Indus Valley traveled in both directions, to Europe and to central and eastern Asia. Along with bronze founding, the actual forms of products—swords, axes, vessels—spread east and west, appearing simultaneously among archaeological finds in the Hallstatt civilization of central Europe and in Shang dynasty China (beginning c. 1600 B.C.).[1]

For medieval Europe, by far the most important source of borrowed technology was China. The most isolated of the ancient civilizations, with its own characteristic cultural patterns and style of thought, China developed its science independent of Greek influence. It remained almost entirely in China. In the early Middle Ages some exchange took place with India in medicine, mathematics, astronomy, and alchemy, but very little with Europe. Joseph Needham suggests that in the medieval world science was everywhere too intimately related to the cultural environment for easy exchange of scientific ideas.

With technology, the case was different. Throughout the first thirteen centuries of the Christian era, technical innovations filtered slowly but steadily from the advanced East to the backward West. "Inventions of immediate practical value" tended to travel "rather than speculations and theories," Needham observes.[2] Some Chinese innovations passed quickly to the West; some took centuries. The course of transmission of some can be tracked step by step; that of others remains speculative.

Transmission Routes of Chinese Technology

One agency of contact between China and the West that became a catalyst for the transmission of technology was silk, the luxury textile par excellence of the ancient and medieval worlds. So closely did the Romans identify the product with its producers that they called the Chinese the "Seres," from the Chinese word for silk.[3] Carried at first through Central Asia over the four-thousand-mile Silk Road and later by sea, silk provided the single permanent line of communication between East and West for many centuries.

According to legend, the mulberry silkworm native to China and the Himalayas was domesticated in the third millennium B.C. when Lei-Tsu, chief concubine of the mythical "Yellow Emperor" Huang-Ti, invented the basic techniques of preparation and weaving while watching a silkworm spin its thread. Whatever the truth of the legend, methods of dealing with the delicate cocoons were devised: soaking in boiling water to kill the chrysalis and remove the gum that holds the fibers together, then picking up the loose ends and reeling the long filaments, "throwing" them (twisting them together), and finally warping and weaving. Eventually silk technology influenced the manufacture of other textiles.

The almost magical attraction of silk was owing to more than its scarcity and cost. The special luster and feel of the fabric—its "hand"—caused by the absence of surface irregularities, gave it the unique quality of "silkiness" that made it a prime article in East-West commerce for more than a millennium.

Silk culture eventually spread to Japan, India, and Persia. The Asian monopoly on silk making was temporarily broken in the fourth century B.C., when Alexander the Great brought the process to the Mediterranean from India. Aristotle described "loosening" threads and "reeling them up" from cocoons that were probably imported rather than cultivated, since the manufacture ceased shortly after. The techniques were forgotten by the time of Pliny (first century A.D.), who believed that silk fibers grew on trees.[4]

The first caravans from China to Persia began to move in 106 B.C., proceeding through either Samarkand or Bactria to Merv, south of the Aral Sea, thence into what is now Iraq, and eventually to the eastern frontier of the Roman Empire. The difficult stage from Iraq to Antioch and Egypt was sometimes avoided by circumnavigating Arabia via the Persian Gulf and the Red Sea, a course that in the first three centuries A.D. became part of a sea route between Greece and India. Later the sea route was extended by Chinese ships that by the fifth century A.D. were sailing as far as Aden in Arabia and the mouth of the Euphrates in Iraq. The sea route gained importance in A.D. 751, when two expanding empires, Chinese and Islamic, clashed in battle at the Talas River in central Asia. Chinese defeat effectively closed the land route for four hundred years.[5]

The sequel to the battle of Talas demonstrates that other groups besides merchants were carriers of technology. A Chinese officer taken prisoner returned home eleven years later to report that Chinese artisans in silk making and other crafts, apparently also captives, had settled at Kufah, in Iraq, the capital of the Abbasid Caliphate. The officer's story, recorded by his brother, the scholar Tu Yu, stated: "As for the weavers who make light silks [in Kufah], the goldsmiths [who work] gold and silver [there], and the painters, [the arts which they practice] were started by Chinese technicians. For example, for painting, Fan Shu and Liu Tzhu from the capital [Sian], and for silk throwing and weaving, Yueh Huan and Lu Li from Shensi."[6] Part of the transmission was the introduction of the horizontal treadle silk loom, probable ancestor of the loom introduced to Europe in the twelfth century for all fabrics.

Other carriers of Chinese technology included émigrés and deserters, like those from the retinue of a Chinese embassy who settled in Persia as early as 100 B.C. and taught their new neighbors "to cast weapons and utensils" of iron. Needham speculates that at about the same time a deserter from the Chinese army taught the people of Ferghana, east of Samarkand, the art of deep drilling, using a technique developed in China a century earlier and due eventually to reach the West.[7]

In the twelfth century, when the Tartars besieged the Sung capital of Khaifeng, they demanded as hostages all kinds of Chinese artisans, including metalworkers, weavers, and tailors.[8] Still other possible agents of transmission were the Buddhist pilgrims traveling between China and India in the early Middle Ages; their interests, however, were less technological than scientific: astrology, mineralogy, and medicine.

Technology as Government Enterprise

Throughout antiquity and the Middle Ages, the direction of technology transmission remained almost entirely from East to West. Europe had little to offer Asia; Asia, and especially China, had much to offer the West. Needham attributes Chinese technological leadership to the strong bureaucratic government that ruled China beginning with the country's first unification under the Ch'in dynasty in the third century B.C. Most of China's engineers and master mechanics were either directly employed or closely supervised by administrative authorities. Complex machines, such as the early water mill, could be developed only under government sponsorship.[9] The same was true of large-scale engineering projects, of which the most famous was the Great Wall, built as a defensive barrier against the nomad horsemen of the North. Extending fifteen hundred miles from the Yellow Sea to a point in Central Asia, the massive project was completed in the third century B.C. during the Ch'in dynasty by joining together walls built earlier by the feudal states.

The identities of many ancient and medieval Chinese engineers and inventors are known. Chang Heng built the first seismograph about A.D. 130 and was the first to apply motive power to the rotation of astronomical instruments. Tu Shih created a water-powered metallurgical bellows in 31 B.C. Ma Chun (fl. A.D. 260) improved the draw loom and invented the square-pallet chain pump, widely used in China and neighboring countries. Yuwen Khai (fl. A.D. 600) superintended the building of the Grand Canal. Li Kao, Prince of Tshao (fl. A.D. 784), developed the human-powered treadmill paddle-wheel warship. The "Leonardo-like figure" Yen Su (fl. A.D. 1030), among many other inventions, designed an important type of clepsydra.[10]

China's large-scale hydraulic engineering projects provided irrigation, water conservation, flood protection, and tax-grain transportation. In addition, as in the Roman Empire, they became a powerful centralizing agency, physically cutting through the boundaries of private estates and asserting superior authority over local governments. The state also built bridges, monopolized iron and steel production, and maintained imperial workshops for the crafts. The Thang dynasty (A.D. 618–906) sustained eight workshops and a Bureau for Barter with Foreign Peoples—an export-sales department.[11]

Imperial interests stimulated innovation in other areas. The practice of geomancy, the pseudoscience that studied "local currents of the cosmic breath" in order to harmonize the palaces and tombs of the imperial family with the influences of the earth, produced discoveries that led eventually to the magnetic compass.[12] Problems of dynastic succession stimulated accurate calendar making and timekeeping by the imperial bureau of astronomy. The government's standardized texts for civil service examinations helped create the demand that presently inspired the invention of printing.

Despite the large government role, however, small-scale private enterprise flourished. Most crafts were worked by individual craftsmen and their families, with skills sometimes concentrated in particular localities—lacquer making in Fuchow, pottery in Ching-te-chen, well drilling in Tzu-liu-ching.[13] Both patterns—family production and the local concentration of skills—were repeated in Europe.

Exporting Chinese Technology

Beginning with cast iron in the fourth century B.C., one major innovation after another appears in the Chinese record, to emerge later, usually only after a long lapse, in the Near East and Europe. In some cases transmission of a device or process from China to the West can be demonstrated; in others, it can only be conjectured, raising the unresolved issue of independent invention, or the possibility of what Needham calls "stimulus diffusion"—the transmission of a general idea, to be developed in different detail. Perhaps the best-known example of such a question is the invention of movable metal type, which took place in China and Korea only slightly earlier than in Europe, and in which transmission is difficult to establish.[14]

An important technology in which China anticipated the rest of the world by at least a thousand years was the production of cast iron. Although China's discovery of iron (513 B.C.) came long after the West's, China leapfrogged ahead in producing cast iron. Chinese ore had a high phosphorus content, which gave it a lower melting point, but the application of waterpower and efficient bellows was probably the decisive factor. By the fourth century B.C., the technique of melting iron and casting it in molds was in regular use for tools and weapons.[15] An ironworks of the Early Han period (second and first centuries B.C.) excavated in Honan contained seventeen smelting furnaces, eight of them blast furnaces, not developed in Europe until a millennium and a half later. The Chinese iron process may have been introduced to the West through the intermediary of Persia in the thirteenth century, though no hard evidence exists.[16]

Waterpower was applied to the iron-casting process by A.D. 31,

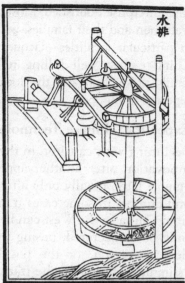

Metallurgical bellows, powered by horizontal waterwheel, from a Chinese work of 1313. [From Joseph Needham, Science and Civilization in China, *Cambridge University Press.*]

when a text records that the engineer Tu Shih "invented a water-powered reciprocator for the casting of [iron] agricultural implements." Smelters and casters were "instructed to use the rushing of the water to operate" their bellows.[17] Waterpower was also applied at an early date to the grinding of grain. The large rotary mill appeared in China at about the same time as in Europe (second century B.C.), but while in Europe the slave- or donkey-powered "Pompeian" mill emerged first, in China the waterwheel took precedence, and when animal-powered mills appeared in about A.D. 175, they were called "dry water mills."[18] As in Europe, mills were highly profitable, both to the great monasteries and to entrepreneurs. Early in the Thang dynasty, the water mill spread to Korea, Japan, and Tibet.

Chinese waterwheels, like those on the periphery of Europe and everywhere east of Persia, were typically horizontal. The vertical wheel was known, however, and used to operate trip-hammers, a single large wheel often turning several shafts;

widespread in China by the third and fourth centuries A.D., it was used not only in forges but in hulling rice and crushing ore. The vertical fall of water was also used in other devices. The spoon tilt hammer consisted of a lever with a hammer at one end and a trough at the other; as the trough filled, the hammer was raised; the trough then automatically emptied, dropping the hammer.[19] The edge-runner mill was another crushing device, in which materials were pulverized by a disk on edge running around a lower millstone. The edge runner appeared in China in the fifth century A.D.[20] Both trip-hammer and edge runner arrived in Europe in the twelfth century.

Floating mills like those of Belisarius were known in China at least by the eighth century A.D. and perhaps much earlier. They may have inspired another invention: the paddle-wheel boat. Instead of water turning a wheel on an anchored boat, the wheel, turned by a man-powered treadmill, propelled the boat. It was probably such boats that astonished the enemy in a naval action in A.D. 418: they "saw the ships advancing up the Wei [River] but could not see anyone on board making them move" because "the men propelling the boats were all [hidden] inside the vessels." Chinese treadmill-paddle-wheel ships reached their apogee in the early twelfth century when a Sung-dynasty marine engineer built a ship carrying a crew of two or three hundred men, with eleven paddle wheels on either side and a stern wheel. The idea eventually reached the West, but much later.[21]

An even more remarkable Chinese application of the water-wheel was its employment in driving mechanized astronomical instruments and finally timekeepers. A succession of planetary models, armillary spheres, and mechanically rotated star maps culminated in A.D. 1090 in a forty-foot-high tower clock built in Khaifeng, capital of the Northern Sung dynasty, antedating the European mechanical clock by more than two centuries. Among the intermediate devices, one instrument, the "Water-Driven Spherical Bird's-Eye-View Map of the Heavens," built in the Thang capital of Chhang-an in A.D. 725, incorporated what seems to have been the world's first escapement, a combination, in the words of a contemporary, of "hooks, pins, and interlocking rods, coupling devices, and locks checking mutually."

The Khaifeng clock of 1090 was the creation of Su Sung, who first built a wooden pilot model, then cast his working parts in bronze. The water that supplied power was contained in a reservoir, refilled periodically by manually operated norias. Water passed by siphon from the reservoir to a constant-level tank and thence to the scoops of the waterwheel. An endless-

Su Sung's astronomical clock-tower, built in 1090 at Khaifeng, was driven by a waterwheel with a complex escapement. Drawing by John Christiansen from Su Sung's description. [From Joseph Needham, Science and Civilization in China, *Cambridge University Press.]*

chain drive slowly turned a celestial globe and an armillary sphere one revolution per day. The same waterwheel turned a series of shafts, gears, and wheels working the bells and drums that announced the time (like all early mechanical clocks, Su Sung's had no face). The escapement that was the "soul of the timekeeping machine" and that kept its movement at an even pace was a complex arrangement of balances, counterweights, and locks that divided the flow of the water into equal parts by repeated weighing, automatically dividing the revolution of the wheel into equal intervals.

In 1126, when the Chin Tartars captured Khaifeng, they destroyed Su Sung's clock tower and carried off the clock, along with several families of craftsmen, who set it up in Peking, the Chin capital. The armillary sphere was struck by lightning in 1195 but was repaired. Early in the thirteenth century when the Chin court fled from the Mongols, the emperor's aides proposed that the armillary sphere be melted down, but the emperor could not bring himself to destroy it, and it was left behind. By the time the Mongols made Peking their capital in 1264, the clock was no longer in working condition.

Meanwhile, in the area beyond the Yangtze to which they had withdrawn, Sung engineers tried to build another clock, but the secret of the escapement had been lost. A contemporary reported, "Now it is said that the design is no longer known, even to the descendants of Su Sung himself." Clock makers returned to the clepsydra, and Chinese clock making reached a dead end, receiving its final blow in the fourteenth century when the Ming dynasty captured Peking and destroyed the remains of Su Sung's famous mechanism.[22]

Perhaps because China never gave a name to its timekeeping devices, they were completely forgotten when, in the seventeenth century, Jesuit missionaries brought to China the European mechanical clock, which the Chinese admired as "a new European invention of dazzling ingenuity" (Needham). Yet Chinese escapements, Needham believes, may have provided at least a stimulus diffusion to the European mechanical clock— perhaps no more than the suggestion that a timekeeper had been invented in the East with a device that divided its movement into equal parts, a problem which Western clock makers solved in an entirely different way.[23]

The wheelbarrow, described in Chinese army specifications in A.D. 230, surfaced in Europe in the twelfth or early thirteenth century. The Chinese "single-[wheel] push-barrow" usually had its wheel in the middle of a boxlike structure, bearing the entire load, so that the pusher moved it but did not support it.[24]

The suspension bridge was first given durable form in the sixth century when Chinese engineers employed iron chains to suspend roadways, a technique copied in Tibet but one that appeared in Europe only at the end of the eighteenth century. The segmental arch bridge, built in China shortly after A.D. 600, was imitated in the West in the fourteenth century.[25] Lock gates built under the Sung dynasty (960–1279) followed in Europe in the late fourteenth century.[26]

A Chinese art that the West tried vainly to copy in the Middle Ages was the production of porcelain, the vitrified (glasslike), fine-grained, and usually translucent pottery in latter days associated with the word "china." In the seventh century A.D., Chinese potters discovered that the mineral feldspar could be incorporated into stoneware, resulting in a primitive type of porcelain. The process was perfected in the thirteenth century under the Mongols by mixing china stone, a rock containing feldspar, with kaolin, white china clay, and firing it at extremely high temperatures (up to 1,450°C). During the late Middle Ages, European potters attempted to imitate Chinese porcelain, their efforts eventually culminating in the sixteenth century in an inferior product, soft-paste or artificial porcelain, a mixture of clay and ground glass fired at a lower temperature. True hard-paste porcelain was not produced in Europe until the eighteenth century.[27]

One pattern that was repeated through antiquity and the Middle Ages was the appearance of a device in Greece and Rome, paralleled or closely followed in China, then forgotten in the West but continuing to develop in the East, and finally revived in Europe. One such device was the odometer, known to Vitruvius and Heron of Alexandria, then disappearing until the end of the fifteenth century, when it was depicted by Leonardo da Vinci. In China the mechanism seems to have originated sometime in the first century B.C. as a mechanical toy, a vehicle in imperial processions whose turning wheels activated drums and gongs. The device's value for surveying and mapmaking was soon recognized, and the wheels and gears were arranged to measure distance. An eleventh-century historian described an elaborate model: "It is painted red, with pictures of flowers and birds on the four sides, and constructed in two stories, handsomely adorned with carvings. At the completion of every *li*, the wooden figure of a man in the lower story strikes a drum; at the completion of every ten *li*, the wooden figure in the upper story strikes a bell."[28]

The "south-pointing carriage," a celebrated vehicle peculiar to China dating from the third century A.D., was once erroneously believed to represent a step in the development of the magnetic compass. A two-wheeled horse-drawn chariot on which a figure was mounted with its arm preset to point due south, the vehicle had gears so arranged that whichever way it turned, the figure pivoted to hold its south-pointing posture. Although it was an invention without further application, whose secret was later forgotten in China, the carriage was not without significance in cybernetics, as a pioneer self-regulating mechanism employing negative feedback.[29]

The magnetic compass, however, did originate in China in the early Middle Ages. Both European and Chinese antiquity were aware of the ability of the lodestone (a variety of magnetite) to attract and repel iron, and of its inductive property—the power to magnetize iron, to impart the same attraction and repulsion to it. Discovery of the directive possibilities of the magnet, however, belonged to China, as did the invention of the magnetic needle, to make readings more accurate.

The earliest certain reference to the magnetic compass goes

back to A.D. 83, in the Han dynasty, when a scholar described the "south-controlling spoon" which when thrown on the geomancer's divining board came to rest pointing south (the arrow on a compass may be, and often was, made to point south rather than north). The board used by geomancers to detect the "winds and waters of the earth" consisted of two plates, the lower square, symbolizing the earth, the upper round, symbolizing the heavens. The upper plate, engraved with the compass points, revolved on a central pivot and bore in the middle a representation of the Great Bear. This plate was turned on its axis to follow the annual path of the bear's tail. The spoon, cut from lodestone, also represented the bear. Aside from its role in divination, some form of "south-pointer" was used by Chinese travelers; a reference from the Han dynasty states that "when the people of Cheng go out to collect jade, they carry a south-pointer with them so as not to lose their way."[30]

The compass matured with the development of more accurate instruments, including the magnetized needle, introduced in China in the eighth century A.D. Created by rubbing an iron needle with a magnet, the device was floated on water on a bit of wood or suspended by a silken thread.[31] By at least the ninth century, the Chinese were aware of the principle of magnetic declination—the fact that the compass needle does not point true north (or south), the inclination varying with the meridian where the reading is taken.[32]

Probably because most of China's water travel took place on its network of canals and rivers or along the coast, the compass was slow to be adopted for navigation. Sometime between 850 and 1050, it began to appear aboard ships, the first certain mention in a Chinese text occurring early in the twelfth century, in a reference to events of the late eleventh: "The [seagoing] ship's pilots are acquainted with the configuration of the coasts; at night they steer by the stars, and in the day-time by the sun. In dark weather they look at the south-pointing needle."[33]

Incidental to the final development of the mariner's compass was a device with a number of applications, an eventual one of which was to provide mounting to keep the compass in horizontal equilibrium, independent of the ship's motion. Called in the West the "Cardan suspension" because it was described, much later, by Italian scientist Jerome Cardan, it was known in China from the second century, when an account described an incense burner with "a contrivance of rings which could revolve . . . so that the body of the burner remained constantly level and could [safely] be placed among bedclothes and cushions." Similar portable stoves were known to the Arabs, who probably transmitted the invention to Europe.[34]

Gunpowder appeared in China as early as the ninth century, when the first reference to the mixing of saltpeter, sulfur, and carbonaceous material occurred in a Taoist alchemy book. The first reaction of the inventors was to warn others against it, lest they singe their beards and burn down their laboratories.[35] Contrary to popular belief, however, the Chinese did not limit early use of the invention to fireworks but very quickly incorporated it into military weapons, evolving from about A.D. 950 into sophisticated rockets and guns.[36] It seems almost certain that the secret was transmitted westward, though the route of diffusion has baffled discovery.

One Chinese invention whose passage to Europe can be traced step by step is paper. A felted sheet of fibers formed from a water-suspension process using a sievelike screen as a mold, paper was first manufactured in China sometime before the Christian era and was widely used in the third century A.D., by which time it had already spread beyond Chinese borders.[37]

Paper may have been first produced accidentally during the process of felting—making nonwoven fabric by applying heat, water, and pounding to plant or animal fibers, or shrinking and matting woven rags by the same process. In the Chinese paper industry, rags were soon replaced in high-grade papers by the bark of the paper mulberry. Gradually the product was improved with sizing and dyes, and by the use of molds made of bamboo strips to replace the earlier screens of coarse cloth.[38]

Cheap and light, paper was first used in China not for writing or printing but for applications such as wrapping. Only in the third century A.D. did it completely replace silk, bamboo, and wood as a writing medium. The two writing materials of the ancient West, animal hides (parchment and vellum) and plant leaves (papyrus), were never used for writing in China, where paper came to have a unique importance. Among its hundreds of uses besides writing were cut-out designs, fans, and umbrellas from the third century; clothing, household furnishings, visiting cards, kites, lanterns, napkins, and toilet paper by the fifth or sixth century, playing cards and money by the ninth.

The use of toilet paper was recorded by a sixth-century Chinese scholar who wrote, "Paper on which there are quotations or commentaries from the Five Classics or the names of sages, I dare not use for toilet purposes." In 851 an Arab traveler commented unfavorably on the cleanliness of the Chinese, who did not "wash themselves with water when they have done their necessities; but they only wipe themselves with paper." Toilet paper was made from rice straw, cheap and soft. In 1393 the Bureau of Imperial Supplies recorded the manufacture of 720,000 large sheets for the use of the court and 15,000 sheets, three inches square, light yellow, thick but soft, and perfumed, for the use of the imperial family.[39]

Paper money seems to have originated in the early ninth century when increased business and government transactions encouraged the institution of "flying money," a credit medium rather than a true money, as a way to avoid carrying the weight of metal coins. Originally a private arrangement of merchants, the system was taken over by the government in A.D. 812 and gradually evolved into a true paper currency.[40]

Two centuries before Cassiodorus sang the praises of papyrus, a Chinese scholar wrote a panegyric to paper in rhymed prose:

> Lovely and precious is this material,
> Luxury but at a small price;
> Matter immaculate and pure in its nature
> Embodied in beauty with elegance incarnate,
> Truly it pleases men of letters.

It makes new substance out of rags,
Open it stretches,
Closed it rolls up,
Contracting, expanding,
Secreting, expounding.

To kinship and friendship scattered afar,
When you are lonely and no one is by,
You take brush to write on paper . . . .[41]

Transmission of paper westward occurred in two stages, the paper and paper products arriving first, followed a century or two later by the manufacturing technique. Neighboring Korea, Japan, and Indo-China learned papermaking as soon as they began to have contact with China (c. second century A.D.). Moving westward over the Old Silk Road, paper arrived in eastern Turkestan, on the shores of the Caspian Sea, in the third century. Chinese paper craftsmen captured at the battle of Talas River in 751 were brought to Samarkand to found an industry that made "paper of Samarkand" an important article of commerce, leading to the establishment of a mill at Baghdad in about 794.[42] Arab manuscripts written on paper survive from the following century.[43] Paper products finally entered Europe in the tenth century through Muslim Spain, which also became the first European country to develop a paper industry, followed a little later by Muslim Sicily and southern Italy. Waterpower was first applied to the pounding process in Baghdad about 950.[44]

In China paper had many advantages over its alternatives, clumsy bamboo and wood and expensive silk. In Europe it had only the advantage of cheapness over papyrus and parchment, and it was more fragile and perishable. Not until the advent of printing in Europe did paper fulfill its potential, replacing parchment for all but the most permanent records.

The course of transmission of printing is much less easily discerned than that of paper. After a long history of preprinting techniques—seals for stamping, stencils to duplicate designs, inked impressions from stone inscriptions—the Chinese began to use woodblock printing in the seventh century. More practical than movable type for written Chinese, with its thousands of ideograms rather than an alphabet, the woodblock dominated Chinese printing for several centuries. A Persian historian described the reproduction and distribution of woodblock-printed Chinese books: a skilled calligrapher copied the author's text onto wooden tablets; these were corrected by proofreaders before being carved on the tablet surface by expert engravers. The tablets (pages-to-be) were then consecutively numbered and placed in sealed bags. When a copy of the book was wanted, the customer paid a charge fixed by the government, and the tablets were taken out, inked, and imposed on sheets of paper.[45]

Since under the Chinese ideograph system the amount of movable type required a large capital investment and elaborate organization of labor, it was profitable only for large-scale production. In about 1045 an artisan named Pi Sheng formed clay characters "as thin as the edge of a coin," fired them; assembled the type on an iron plate coated with pine resin, wax, and ashes; warmed and cooled the plate to solidify the type; then inked it and made the impression. A contemporary, describing the process, explained, "If one were to print only two or three copies, this method would be neither simple nor easy. But for printing hundreds or thousands of copies, it was marvelously quick." Pi Sheng usually worked with two forms, taking the impression from one while type was being set on the other, enabling the printing to be done "with great rapidity." The type was arranged in wooden cases with paper labels, "one label for words of each rhyme-group."[46]

The first practical wooden movable type appeared late in the thirteenth century, when a magistrate named Wang Chen used characters cut out of wood blocks with a small, fine-toothed saw, then finished with a knife for exact uniformity and arranged for easier handling in compartmental wooden cases on revolving tables.[47] Metal movable type was developed early in the fifteenth century in Korea, where three cast-bronze fonts were made before the appearance of metal type in Europe or China.

A reasonable conjecture is that printing followed the path of paper to Turkestan, whence it reached Persia during the thirteenth-century Mongol domination of central Asia, to appear in Europe, first in block form and then as movable type.[48]

One outstanding Chinese invention that did not migrate westward was the junk, one of the best sailing ships ever designed. Powered by square sails composed of linen panels that were raised and lowered like venetian blinds, the junk had a high stern and massive stern rudder (an idea that did eventually reach the West) that on the junk doubled as a keel. Another notable feature was the watertight compartment.

The Technology of India and Persia

The technological contributions of India and Persia, many of them fundamental inventions, belong mostly to prehistory and antiquity. Indian metallurgy was far ahead of European; the Romans imported Indian steel. Medieval India produced landmark scientific advances, notably in mathematics—algebra and the so-called Hindu-Arabic numerals, embodying the principle of place value and the zero. Of its technical innovations, one, the *churka* (cotton gin) had an enormous impact on the West. India may also have produced the original ancestor of the spinning wheel, for use with cotton.

Persia's chief medieval invention was the Eastern version of the windmill, mentioned for the first time in the seventh century. Like the Eastern waterwheel, it was horizontal, with enclosing walls admitting the wind on one side. Windmills of the Persian type spread to Turkestan and thence to China, where they took on a nautical form, with fore-and-aft sails mounted on masts around a drum. If a connection exists between these horizontal windmills and the vertically mounted windmill that began to appear in Europe in the twelfth century, it is probably that of stimulus diffusion.

The Arabs, Transmitters and Inventors

In the immense transfer of science and technology that marked the Middle Ages in Europe, Africa, and Asia, the Arabs played a unique role. Along with the spices and silks they carried from China and India, they brought many of Asia's discoveries and inventions, and they provided the means by which Europe at last recovered its own lost heritage of Greek knowledge. The protracted conflict between Islam and Christian Europe has obscured the remarkable service performed by the former to the benefit of the latter, but the process occurred naturally enough. "Between the eighth and the twelfth centuries, the sophistication and culture of the Islamic world made it the suitable heir of classical civilization" (Joel Mokyr),[49] and as such it readily absorbed the science and philosophy of Greece.

The Arab Age of Translation began during the reign of Harun-al-Rashid (A.D. 786–809), when scholar-physicians at a Nestorian Christian academy in Jundi-Shapur, in southwest Persia, were brought to Baghdad to translate Greek manuscripts gathered by the caliph's agents, acting, in the words of a modern writer, as "buyers of culture."[50] A young scholar from Jundi-Shapur, Hunayn ibn-Ishaq, became court physician to Harun's son, Caliph al-Mamun, and in 830 was named head of the "House of Wisdom," a library founded by the caliph to store and translate Greek manuscripts. Hunayn and his colleagues translated Plato's Republic, many of Aristotle's works, and the medical writings of Hippocrates, Dioscorides, and Galen (some of whose works were later lost in the original Greek and preserved to the world solely in Hunayn's Arabic).[51]

Most of the Greek works translated belonged to the Hellenistic period, and the culture they represented was not that of literary Athens but that of scientific Alexandria. Homer was never rendered into Arabic, nor were the works of Greek historians or dramatists. Aristotle's comments about Greek drama in the *Poetics* puzzled Arab readers; they had no theater of their own and were totally unacquainted with Sophocles and Euripides. Their interests were essentially those of Aristotle himself: the natural sciences, medicine, chemistry, astronomy, mathematics, geography, and the philosophy underlying them. They did not stop with preservation and translation; Arab scholars expounded and interpreted the Greek material, to the great benefit of later European intellectuals. The most famous of the Arab commentators were Avicenna (Ibn-Sina, 980–1037) and Averroës (Ibn-Rushd, 1126–1198). The distinguished astronomer and mathematician al-Khwarizmi (c. 780–c. 850) revised the geography of Ptolemy and wrote an original treatise expounding the Hindu numerals which, in Latin translation, introduced them to the Christian West.

In other fields the Arabs made original contributions. Commerce was a respectable, even prestigious profession in the

Muslim philosopher Averroës (Ibn-Rushd).
Detail of fresco The Triumph of St. Thomas
Aquinas, *by Andrea da Firenze, in Santa Maria
Novella, Florence. [Alinari.]*

world of Islam. "The honest Muslim merchant will rank with the martyrs of the faith," said Muhammad, and "Merchants are the couriers of the world and the trusted servants of God upon earth."[52] Islamic business techniques in banking, bookkeeping, and coinage were so far in advance of those of Europe that when the Normans conquered Sicily (1071–1091), their pragmatic Christian kings employed Muslims to handle their finances.[53]

To the science of distillation, already centuries old, Muslim alchemists made many practical contributions. Muslim musicians introduced the first bowed instruments, the lute and rebec, ancestors of the violin. Following the First Crusade, European military engineers learned much of the art of castle building from their Muslim foes. The secrets of Syrian glassmaking were sold to Venice (in 1277) and its techniques taught by Muslim artisans, founding a monopoly of the manufacture of fine glass long maintained by that city.[54]

Undergirding the transformation of the Muslim half of the Mediterranean world lay Islam's own revolution in agriculture. Like so much else, the crops and processes of the new system were borrowed from Asia, in this case mostly from India. Muslim enterprise, public as well as state-encouraged private, combined them into a fresh synthesis. A wide array of food and fiber plants was introduced, new patterns of cultivation were adopted, and extensive irrigation systems were built. The new crops included rice, sorghum, hard wheat, sugarcane, cotton, watermelons, eggplants, spinach, artichokes, sour oranges, lemons, limes, bananas, plantains, mangoes, and coconut palms. Most required intensive cultivation with application of fertilizer, heavy watering, and a flexible system of crop rotation that employed all the seasons of the year. New irrigation devices, also borrowed rather than invented by the Arabs, included dams, drainage tunnels, canals, and water-lifting machines. The system with its wider variety of crops, more land under cultivation, and more intensive cropping, helped stabilize Islamic agriculture, stimulating denser rural settlement and the growth of cities.

Like other elements of Islamic civilization, the new agriculture migrated from Baghdad westward, by way of Egypt, Tunisia, and Morocco, to reach Europe via Muslim Spain. Some of the techniques, notably irrigation works, were quickly imitated in Christian Europe, whose southern regions also adopted the cultivation of cotton, rice, sugarcane, and citrus fruits. At the end of the Middle Ages, many of these were successfully transplanted to the Americas.[55]

In stock farming, Spanish Muslims bred the famous merino sheep, which made Spanish wool preeminent in the world. But it was in wool's great rival, cotton, that Islam made its most important textile contribution, transmitting to Europe the secrets of its cultivation and conversion into cloth. The Arabs were the first people in the Near East and Europe to adopt cotton for ordinary clothing. Muhammad himself set the style by wearing a white cotton shirt and trousers under a woolen cloak, a costume adopted by the first caliphs. In later times, rulers and the very wealthy wore silk and embroidered cloth, but everyone else, in both town and country, wore cotton, either white or black—cotton undergarments, cotton caftans and robes, cotton mantles for women, cotton veils and turbans. Cotton was used for shrouds and funeral clothes, and for bedclothes, tablecloths, curtains, towels, and rugs.

Housed in a network of textile workshops known as *tiraz* factories, royal textile manufacturing was closely regulated and workmanship and production were strictly controlled. Alongside the *tiraz* establishments in the great cities, a flourishing private industry was carried on in surrounding smaller towns. The most prized cotton fabrics produced for international commerce came from the *tiraz* cities of Iraq, Persia, and Syria, while the lesser towns produced under private management textiles for local and regional markets. Arab conquest brought cotton manufacture and cultivation to North Africa, Spain, Sicily, and southern Italy. In Spain the Arab cotton industry disseminated technical knowledge via regions reconquered by the Christians. In Sicily and Italy, Europeans inherited the Arab textile systems intact.[56]

* * *

Well into the Middle Ages, Europe remained a poor relation of Asia, accepting hand-me-down technology from China and India, carried westward either in artifact or in idea by intermediaries of whom the most important were the Arabs. Up to the thirteenth century, there was scarcely any direct contact between the two regions; an English cleric, French knight, or Italian merchant knew hardly more of China than had Julius Caesar. Yet through these centuries, Asian technology was steadily infiltrating Europe, unheralded and unrecognized, but with growing impact on the European way of life and patterns of work.

5

THE TECHNOLOGY OF THE COMMERCIAL REVOLUTION

900–1200

A FTER 900 EUROPE'S DEFENSES AGAINST Viking and Saracen marauders gradually stiffened, containing, converting, and absorbing the Scandinavians, and pushing back the unconvertible Muslims by counteroffensive. In Spain, Christian kings slowly gained the upper hand over their Muslim rivals; a band of Norman French adventurers successfully invaded Arab-held Sicily in 1071; and in the last decade of the eleventh century, the religious outburst that fueled the First Crusade carried the European counteroffensive into Asia Minor, Syria, and Palestine.

Whether Christianity gained anything from the capture of Jerusalem is debatable, but the Italian maritime cities, in particular Pisa and Genoa, unquestionably profited from the seizure and colonization of the coastal towns of the Levant. The long-distance commerce in the luxury goods of the Orient carried on through these colonies played an important role in what Robert S. Lopez has called "the Commercial Revolution of the Middle Ages," the economic surge in which the Italian cities were the energetic leaders.

The other military victories over the Muslims brought dividends hardly noticed at the time but of immense future significance: access to Islamic learning and technical knowledge. Toledo, taken by Alfonso VI of Castile in 1105 and converted into his new capital, proved a tremendous cultural prize, the decline of Baghdad having made it the leading center of Islamic learning.[1] Sicily and southern Italy bestowed another treasure of scientific and technical information. The principal European gain from military success was thus not the expulsion of the infidels but the opportunity to mix with them. In Spain the strengthening of Muslim defenses by reinforcements from North Africa prolonged the interface by stabilizing the military and political front for a century.[2]

The Vikings, meanwhile, turned their attention westward, where Irish missionaries had discovered Iceland. Reconnaissance was followed by settlement of the large island, up to then inhabited only by foxes. From Iceland's western shore another island could be discerned some 175 miles away. Eric the Red colonized Greenland in the 980s, and one of the first ships to sail for the new colony, commanded by Bjarni Herjulfson, missed its landfall in the fog and was blown across the Davis Strait to Labrador. Bjarni inspected the forested coast and without attempting to land became, in 986, the first known European discoverer of America. In the millennial year of 1000, Leif Ericsson undertook to follow up the discovery (using Bjarni's own ship) by planting a colony there.[3] As an achievement, Leif's voyage was unimpressive, merely another crossing of the Davis Strait. As a historic event it also proved insignificant. The colony vanished almost without a trace, and though other Viking ships visited North America in the following two centuries (mainly in search of wood for unforested Greenland), they made little impact.

Bjarni and Leif may not even have been Europe's first discoverers of America; they may have been preceded by Irish missionaries, sailing their toylike skin-covered boats via Greenland to Labrador or Newfoundland.[4] As a practical enterprise, any such project was doomed; Europe was not ready to discover America, which it did not yet need. The missionary effort, however, has a certain symbolic significance, as a reminder of the early roots of one of the great motivations of explorers.

In the centuries after Eric, Bjarni, and Leif, Europe was at last getting ready to discover America and to do much more. "From 950 on," writes Robert Reynolds, "there was . . . growing manufacture of textiles, pottery, leather goods, and many other things. The list of articles manufactured gets longer and longer [as the tenth century gives way to the eleventh], the products get better and better. Prices go up in terms of money but down in terms of man hours because of more efficient management, the application of mechanical power, improvement in tools and machinery, and better transport and distribution."[5] Where Europe had formerly exported "low-grade, backward-area" products such as slaves and furs, by Leif Ericsson's time it had begun shipping textiles and metal products to Africa and the Near East, and even to Asia. By 1200 it was sending high-grade woolens to Alexandria, Constantinople, and farther east, as well as bar iron, copper ingots, utensils, and arms and armor. Returning ships carried grain to Europe's cities from Sicily and North Africa. A notable import was chemicals, especially alum, used by dyers to fix colors in the ever-expanding wool cloth industry.

In the north, several German towns situated on or with access to the North Sea—Cologne, Bremen, Hamburg—launched careers as major commercial carriers. By the year 1000 their merchants had won a privileged position in London for their "Hanse," or merchant guild, in time gaining virtual exemption from customs by lending money to Richard Lionheart.[6]

In a word, Europe was turning from a developing into a developed region. The growth of industry meant the growth of cities, which in the eleventh and twelfth centuries began to abandon their old roles of military headquarters and administrative centers as they filled with the life of commerce and industry. Some, like Genoa, once Roman villages, mushroomed, while others, like Venice, appeared out of nowhere. Still others, calling themselves simply "New City" (Villanova,

Villeneuve, Neustadt), were founded by progressive rulers. Instead of growing haphazardly, they were built on a plan, typically a grid pattern with a central square, church, and market buildings.[7] Beginning in tenth-century Italy, businessmen and craftsmen in many cities established what they called "communes," declaring themselves free men who owed allegiance only to a sovereign who collected taxes but otherwise left them alone. Astute lords granted charters exempting city dwellers from feudal obligations—"so that my friends and subjects, the inhabitants of my town of Binarville, stay more willingly there," sensibly explained one lord.[8] Under the rubric "Free air makes free men," even serfs were declared emancipated if they maintained themselves in a city for a year and a day.

A central feature of the Commercial Revolution was the trade fair. Dating back to Roman times, the fair was kept alive through the early Middle Ages by maritime centers where native and foreign merchants could conveniently meet. The practice spread to inland towns in Italy and southern France; in northern France, St. Denis, near Paris, opened a highly successful annual fair in about 635. In the ninth century, many more were inaugurated, spreading in the tenth to Flanders and northern Germany. But by far the most famous and significant were the Fairs of Champagne, the region east of Paris. In the eleventh and twelfth centuries, the counts of Champagne organized them into an annual cycle of six fairs occupying the entire year, with safe conduct guaranteed to foreign merchants. The Champagne Fairs forthwith became the established rendezvous for merchants from Italy and Flanders.

The Middle Ages had inherited no effective credit instruments from the ancient world, which had never developed any. Furthermore, the Christian Church, in accordance with the Bible, condemned usury and defined it as the charging of any interest whatsoever. The revival of commerce and the transactions of the Champagne Fairs stimulated the invention of novel forms of credit designed to circumvent the Church's ban. Italian merchants had already adopted one form, from Arabs, Jews, or Byzantines, the *commenda* (also called the *collegantia* or *soci-*

etas), by which one partner undertook a voyage, borrowing all or most of the capital from the other partner, with profits divided according to a prearranged formula—an insurance device as well as a loan and a temporary partnership. Another form of credit developed simultaneously was the *cambium maritimum*, in which a merchant undertaking an ocean voyage borrowed capital in one place in one currency, to be repaid in another place in another currency, the interest concealed in the rate of exchange, repayment contingent on safe arrival of the ship. The Champagne Fairs converted this form of loan into a land arrangement, omitting the element of marine insurance and creating both a way to earn interest and a means of transferring funds abroad, thus making possible the transaction of business by remote control.[9]

Along with the cities, the farming regions that fed them grew. A European population between the Baltic and Mediterranean estimated at 27 million in A.D. 700 had reached 60 or 70 million by 1200.[10] In Carl Stephenson's words, the creation of "a bourgeois class such as the Roman Empire had never seen" gradually brought with it "the emancipation of the rural masses that made possible our modern nations."[11]

Behind this demographic and economic surge lay technical innovations: a radically new system of organizing agricultural work, newly expanded power sources, dramatic new techniques in building construction, and other novelties undreamed of by Greeks and Romans.

Open Fields and Waterpower

By the central Middle Ages, much of the countryside was dominated by two complementary systems: on the level of the village and its peasant inhabitants, the form of organization of work known as the "open-field system"; on the level of the lord, the form of management called the "manorial system."

The manor, often not coincidental with the village (which might contain more than one manor, or be only a part of a manor), was an estate held by a lord. In its classic form it con-

Snow highlights house sites, gardens behind them, and surrounding fields in aerial photograph of deserted open-field village of Wharram Percy, Yorkshire. [Cambridge University Collection of Air Photographs.]

sisted of land directly exploited by the lord (the demesne) and peasant holdings from which he collected rents and fees, usually including labor services. The combination of demesne and tenants probably dates from the early Middle Ages, but it had its first specific documentary mention in the ninth century in northern France and in the tenth century in Italy and England. By the eleventh century it was well established in Europe.[12]

The origins of open-field agriculture are lost in the obscurity of the tenth and eleventh centuries, whose scanty documentation offers only scattered scraps of information. By the time the more abundant records of the twelfth century provide light, the system was already mature.

Several factors may have contributed to this second agricultural revolution. One was population growth, fragmenting family holdings through the custom of dividing inheritance among children or among sons. A second was "assarting," cultivating new land or reclaiming wasteland. When a group of peasants banded together to clear forest or drain swamp, they divided the resulting "assart" into strips convenient for plowing. A third factor was the heavy plow, which favored working long strips over square plots, to reduce the number of turnarounds, especially awkward with multiple-animal teams. In time, broad areas of Britain and continental Europe were occupied by villages surrounded by two or three large fields made up of cultivated strips clustered in "furlongs" fitted to the contours of the terrain. Each year one field was left fallow, the remaining land cultivated with spring and fall crops. Crop rotation was biennial in a two-field system, triennial in a three-field.

The strips were not apportioned equally; some peasants held several, some few, some none. The lord's demesne usually consisted of a large number of strips, cultivated by the serfs (villeins, in England) in payment for their holdings. Every tenant, free or serf, who held strips held them in both or all three fields, to ensure a crop every year.

Crop rotation and fallow were not new devices. What was new was the way they were organized and regulated. Plowing and planting, harvesting and opening the fields to grazing, all had to be done in concert. The necessary decisions were made not by the lord but by the peasants, whose cooperation became the hallmark of the system. While meeting their obligations of labor and money payments to the lord, the villagers created their own self-governing apparatus based on a set of bylaws that ruled their working lives. The system was neither free enterprise nor socialism; it was sui generis, one of the unique creations of the Middle Ages.[13]

The heavy plow and the new horse harness fitted well into the open-field system, even if the two devices probably did not, as was once believed, play the decisive role in its establishment. The plow heightened the need for cooperation, since not all peasants who held land owned plows or plow animals.

Expanded use of the horse stimulated cultivation of oats, a spring crop appropriate to open-field rotation. The fact that the horse was fed in the barn made his manure easy to collect and thus increased the use of fertilizer, while the spring legumes (peas, beans, and vetches) restored nitrogen content to the soil. Another fortuitous benefit of the system emerged in the ridge-and-furrow pattern built up by strip plowing. In the wet but erratic climate of northern Europe, the ridge tended to stay dry in wet seasons and the furrow to stay moist in dry seasons, providing a kind of crop insurance.

Not all medieval agriculture was open-field. In some regions isolated homesteads were the rule. Other systems existed, notably the infield-outfield system, in which a small "infield" was worked intensively with the aid of fertilizer while the large "outfield" was held as a land reserve.

The population surge stimulated another response in the Low Countries, where an unending war against the sea had been carried on since the seventh century. The Frisians and other coastal dwellers depended for life and livelihood on precarious seawalls four or five feet high, which they maintained with great difficulty until the monks of the new Cistercian Order came to the rescue. An arrangement was worked out by which the land was deeded to the monasteries, then leased back to its cultivators while the monks took responsibility for upkeep of the dikes. Lay brothers performed the labor despite recurrent overwhelming floods. A chronicler of one such flood left an arresting image: a floating tree, to which clung a man, a wolf, a dog, and a rabbit.[14]

Another new monastic order, the Carthusians, dug the West's first deep-drilled well in 1126, at Lillers, in Artois (whence the name "artesian" well). A shaft only a few inches in diameter was sunk through impermeable strata to reach a stratum of water under pressure, producing a well that needed no pumping. The technique—percussion drilling, a succession of blows struck on a rod with a drilling tool on its end—had long been used in China. Whether it was borrowed or independently invented in Europe is conjectural.

The tenth and following centuries witnessed steady progress in reclamation of unproductive areas via drainage, irrigation, and land clearance. Northern and western Europe, once sparsely inhabited, filled in. By the end of the twelfth century, the fields, meadows, and woodland of thousands of villages abutted one another. All of them cultivated grain, and most ground it by water mill. The rapidly multiplying written records supply a wealth of statistics, of which perhaps the most telling is the figure given in Domesday Book, the survey prepared in England in 1086 at the order of William the Conqueror. A century earlier, fewer than 100 mills are recorded in the country; Domesday Book lists 5,624 (low, since the book is incomplete). Georges Duby calculates that the figure indicates a mill for every forty-six peasant households and points out an implication: a substantial rise in consumption of baked bread in place of boiled, unground porridge.[15] Continental records tell a similar story. In one district of France (Aube), 14 mills operated in the eleventh century, 60 in the twelfth, and nearly 200 in the thirteenth.[16] In Picardy, 40 mills in 1080 grew to 245 by 1175.[17] The boat mills invented by Belisarius, moored under the bridges of early medieval Paris and other cities, began in the twelfth century to give way to structures permanently joined to the bridges.

The waterwheel never played a major role in the Muslim world, not for lack of knowledgeability—Muslim hydraulic engineering was far ahead of European—but for want of fast-

Water mill with overshot wheel and eel trap in the millstream. [British Library, Luttrell Psalter, Ms. Add. 42130, f. 181.]

flowing streams. Large dams and intricate irrigation systems aided agriculture in Moorish Spain, but the waterwheel was used only for grinding grain and raising water.[18] In Christian Europe, in contrast, the vertical wheel, including the powerful overshot type, was finding important new applications. Once more the monasteries led the way. The great Benedictine abbey of St. Gall in Switzerland pioneered the use of waterpower for pounding beer mash as early as 900.[19] The new Cistercian reform movement launched in 1098 at Cîteaux, in Burgundy, carried on the Benedictine tradition of promoting technology by founding waterpowered grain mills, cloth-fulling mills, cable-twisting machinery, iron forges and furnaces (where the wheels powered the bellows), winepresses, breweries, and glass-works. The edge-runner mill, long known to China, was adopted for more efficient pressing of olives, oak galls and bark for tannin, and other substances requiring crushing.[20]

The contemporary biographer of St. Bernard, leader of the Cistercian movement, illustrated the respect accorded the waterwheel; in describing the reconstruction of the saint's abbey of Clairvaux in 1136, he neglected the new church but included an enthusiastic account of the monastery's waterpowered machines.[21] The first waterpowered iron mills in Germany, England, Denmark, and southern Italy were all Cistercian.[22]

One of the earliest widespread industrial applications of the waterwheel was in fulling cloth; the trampling feet of the fullers were replaced by heavy wooden hammers lifted and dropped by the turning waterwheel. One effect was to draw the fullers into the countryside, where they further profited by freedom from the sometimes restrictive regulations of the towns. Another effect was the spread of the knowledge of gearing.[23]

Hemp production required a similar pummeling action to break up the woody tissues of the dried stalks and free the fibers for manufacture of ropes and cords. The existence of a water-powered hemp mill is documented in the Dauphiné, in south-eastern France, as early as 900.[24]

By the late eleventh century, waterpower was pounding, lifting, grinding, and pressing in locations from Spain to central Europe. In several applications of waterpower, notably in lifting and dropping hammers, the camshaft made its earliest Western appearance, diffused from China (as Joseph Needham believes) or independently invented, as seems not unlikely. The cam, a small projection on the horizontal shaft of a vertical water-wheel, caught and lifted the falling hammer, which dropped of its own weight. Usually a pair or more of cams on the same shaft operated alternately.[25]

The more abundant records of the twelfth century throw scarcely more light than the scanty ones of the eleventh on the types of waterwheel that were built, but increasingly the efficient vertical overshot wheel justified its initial cost when used to grind grain, and its superiority was persuasive in industrial applications. Modern calculations show that the ancient donkey- or slave-powered quern of Rome produced about one half horse-power, the horizontal wheel slightly more, the undershot vertical about three horsepower, and the medieval overshot wheel as much as forty to sixty.[26] A continuing weakness of waterwheel installations was their reliance on fragile wooden parts, gears as well as camshafts. On the other hand, a broken wooden piece was easily replaced by a peasant craftsman. The cost of iron made metal gears a luxury even as late as the eighteenth century.

The proliferation of gristmills may have owed as much to the role that they had come to play in the lord-peasant relationship as to their labor-saving value. An aspect of the "ban," the lord's local power, became one of the most resented of his privileges: the obligation of unfree peasants to have their grain ground at the lord's mill, at the cost of a multure, commonly one thir-teenth of the grain or flour. Despite peasant protests, the prof-itable ban was gradually extended to oil and wine presses, bake ovens, and iron forges. The peasants' resentment was not only against the payment of multure but against the inconvenience of taking the grain to the mill and waiting in turn. The right to "jump the queue" was reserved to free tenants willing to pay a fine to grind "next after the grain which is in the hopper." Many peasants surreptitiously operated illegal hand querns, while others went back to eating porridge.

The gristmill often represented the most visible symbol of a wider oppression, as in the rebellion of the inhabitants of

St. Albans. Seeking to secure a charter of urban liberties from the lord abbot, the townspeople openly and defiantly milled their own grain with their hand querns. When the rising was suppressed, the abbey confiscated the querns and incorporated the millstones into the floor of the monks' parlor as a trophy of victory. Fifty years later, during the Peasants' Rebellion of 1381, the people of St. Albans dug up the monks' floor and distributed the fragments of the stones among themselves as tokens of solidarity, in the spirit, according to St. Albans chronicler Thomas Walsingham, of sharing the sacrament. Struggles over hand mills in other places likewise signaled deeper grievances about taxes, labor services, and legal status.

Some scholars have argued that the lord's gristmill was economically viable only because of the ban and that without the ban the tenants would have chosen to use their own cheaper and more convenient hand mills. Evidence to the contrary, however, is found in the existence of a special rate of multure charged to free tenants, who were not under obligation to use the lord's mill. This rate, typically one twenty-fourth, a little more than half that charged unfree tenants, seems to represent the true, free-market value of the service. Recent scholarship has also discovered the presence of independent water mills, outside the lord's control and competing with his mills, held by free tenants and even by villeins, who charged their own multure fees and kept the profits. Such mills seem sufficient proof that compulsion was not the only basis for the gristmill and that powered milling of grain made compelling economic sense even without the ban.[27]

Waterpower spurred construction of dams, at first on a small scale to create millponds and millraces but increasingly on a larger scale. The Arabs, who in their era of conquest had learned about dam building from India and the Near East, brought their knowledge to Spain, where a few Roman dams still operated. These they kept in repair, adding dams of their own, such as the great structure at Murcia, 425 feet long and 25 feet high, its rubble core faced with masonry blocks.[28] By the twelfth century, dam building had crossed the Pyrenees in a spectacular form. At Toulouse, forty-five mills were driven by streams controlled by three dams in the Garonne. The principal one, mentioned in a document of 1177, was probably the largest dam then existing. Thirteen hundred feet long, it was built diagonally across the river by ramming thousands of giant oak piles into the riverbed to form palisades that were then filled with earth and stone.[29]

Millraces similarly expanded into hydropower canals in the twelfth century. The monastery of Clairvaux dug a 3.5-kilometer (2-mile) millrace canal from the river Aube to the abbey, while the Cistercians of Obazine chipped one 1.5 kilometers through solid rock.[30]

Medieval engineers were the first to exploit the waterpower supplied by ocean tides. Tidal mills are recorded in Ireland as early as the seventh century, in the Venetian lagoon before 1050, near Dover in Domesday Book, and a little later in Brittany and on the Bay of Biscay. The practical value of tidal mills was limited by their short operating periods (six to ten hours a day), the eccentric working hours imposed on the millers, and the vulnerability of the mills to storm damage.[31]

In the last twenty years of the twelfth century, an entirely new prime mover appeared simultaneously on both sides of the English Channel and the North Sea. Nothing like the windmill in its vertical European form had ever been seen. Though some scholars believe it to have derived from the horizontal windmill of Persia, perhaps diffused through Muslim Spain, the weight of evidence favors an independent origin, possibly in East Anglia, where it replaced unsatisfactory tidal mills and supplemented the scanty waterwheels. Reversing the waterwheel's arrangement, the windmill placed the horizontal axle at the top of the structure, to be turned by sails, gearing it to the millstones below. The immediate problem of keeping the sails faced into the wind (or out of it in a gale) was solved by balancing the mill on a stout upright post, on which it could be turned, none too easily, by several sturdy peasants gripping a large boom.

Crafts in Town and Country

While agriculture still absorbed much the largest share of human labor, the number of handicraft workers and the volume and diversity of handicraft production increased significantly in the central Middle Ages. By far the most universal and widespread craft was that of cloth making, which experienced a major change in the organization of work. Throughout antiquity and the early Middle Ages it had been the province of women working in the gynaeceum; now, as slavery waned and the gynaeceum disappeared, the craft was decentralized to the family unit. The male head of the household was the weaver; the women prepared and spun the yarn for his loom.[32]

By the twelfth century, the loom the weaver operated was a

A fourteenth-century windmill.
[British Library, Stowe Ms. 17, f. 89v.]

new mechanized model, probably a descendant of the Chinese silk loom. Horizontal rather than vertical, it allowed the weaver to sit while he worked. In operating the old vertical warp-weighted loom, the weaver, standing, had advanced or retarded sets of warp threads by moving the heddle bar forward or backward in its brackets; now this operation was performed by a pair of foot treadles, leaving the operator's hands free. To pass the weft yarn through the shed created by the heddles, he used another innovation, the boat-shaped shuttle, holding a bobbin wound with thread.[33]

To the English scholar Alexander Neckam (1157–1217), the weaver at the horizontal loom was "a horseman on terra

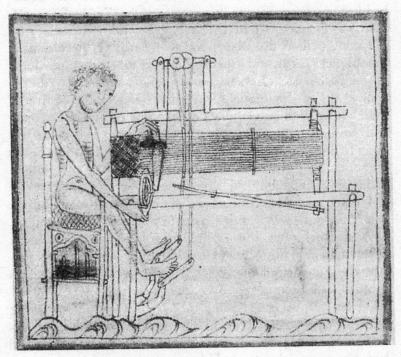

Horizontal loom, c. 1250. The weaver "leans upon two stirrups" (the treadles, which control the heddles by means of an overhead pulley arrangement), while he passes back and forth the newly invented boat-shaped shuttle. (The plane of the warp threads is parallel to the floor, despite the faulty perspective.) [Trinity College, Cambridge, Ms. 0.9.34, f. 34v.]

firma who leans upon two stirrups," the treadles, which were attached to a pulley arrangement above the loom called the "harness."[34] The shuttle was weighted and contained "an iron or wooden bobbin," from which the weft thread was drawn as "the one hand of the weaver tosses the shuttle to the other, to be returned vice versa." As he completed a row, the weaver "beat down the work accomplished." Alexander pictured a woman, a *textrix*, working alongside the weaver, combing the wool, spinning the yarn, smoothing the finished cloth, and helping the time pass by singing "sweet songs" to the man at the loom.[35]

From about 1000, wool cloth manufacture for the market had begun to concentrate in certain regions. Especially prominent was Flanders, where the favorable conditions included a soil congenial to dye plants, an abundance of the cleansing agent fuller's earth, and proximity to England, the prime source of fine fleece. As in Roman times, fulling and dyeing were specialized trades. Traditionally, the fuller trampled the cloth in a trough filled with water, fuller's earth, urine or wine dregs, lime, and sand, changing the water several times. The cloth was hung to dry on a wooden frame, the "tenter," fastened by hooks that could be adjusted to stretch the fabric to the right length and breadth and then teaseled to raise the nap.

Dyeing might be done at different stages in manufacture: before the yarn was woven ("dyed in the wool"), after fulling, or sometimes after the cloth was sold. Among the most demanding of cloth techniques, it required a knowledge of fabrics, dyes, and mordants (color fixatives) and by the twelfth century had been taken over almost completely by men.[36] Cloth or yarn was soaked in a tub of hot water and turned at intervals with a pole. Red and blue were the predominant colors, produced respectively from the madder root and the woad leaf. Since woad leaves could be plucked several times a year, blue dye was cheaper.[37]

The wool cloth cities of the eleventh and twelfth centuries were commercial rather than industrial centers, collecting cloth from neighboring villages to be sold in Baltic and

The boy Jesus as an apprentice dyer, c. 1300. [Bodleian Library, Ms. Selden Supra 38, f. 27.]

Mediterranean markets.[38] The entrepreneurs who carried on this trade formed associations called "merchant guilds," first in tenth-century Italy. The origins of the merchant guilds were in religious and charitable brotherhoods and "associations of fellow drinkers" (Robert Lopez), but they became powerful monopolies which in many cities dominated the political as well as the economic scene.[39] In Florence members of the *Arte di Calimala*, who clustered on the Via Calimala, bought unfinished cloth at the Champagne Fairs in France and transformed it into a luxury commodity by their superior finishing and dyeing. Their dyes were imported and expensive: red from "kermes" or "grana," derived from certain dried insects; violet from orchil, a lichen; crimson and purple from brazilwood, an East Indian tree. Their "scarlets," particularly prized, were not necessarily red but sometimes gray, dark gray, dark blue, or black, the common element a certain proportion of kermes. Costing as much per length as a skilled mason could make in three years, the scarlets were for the wealthy elite: "popes, emperors, archbishops, princes—and . . . Flemish civic aldermen" (John Munro).[40]

Wool dominated the textile industry, but linen retained importance, based on a technology long known to the West. At first practiced in the towns, linen making was shunted to the outskirts or the countryside when the retting ponds, where the woody flax tissues were soaked to separate them from the fibers, became a public nuisance rivaling the noisome operations of the tanners and butchers.

By the twelfth century, cotton and silk were also being manufactured on a substantial scale in Italy. Sicily's cotton industry, inherited by the Norman conquerors from the Arabs, was supplemented in the early twelfth century by new centers of manufacture in northern Italy. Both products and techniques imitated those of the Arabs, with two differences. Where the Arab industry was largely government controlled, the Italian remained in the hands of private entrepreneurs, and where the Arab industry stretched from Baghdad to Spain, near its widely scattered sources of raw materials, the Italian was concentrated in the Po plain, whither it imported raw cotton from overseas and where it soon became for European cotton what Flanders was for wool.

Italian craftsmen ginned their cotton with the Indian churka, acquired from the Arabs, a device not improved on until Eli Whitney's invention. It consisted of two grooved wooden rollers, turned with a crank, revolved against each other in opposite directions to remove the seeds. The bolls were beaten, an operation facilitated by the introduction in the early twelfth century of the *arco*, a wooden bow suspended from wall or ceiling, its taut cord buried in a pile of raw cotton. When the cord was tapped with a mallet, its vibrations caused the cotton bolls to open and the fibers to separate. The cotton was combed or carded and spun, and half of it "warped," grouped into uniform lengths to be positioned on the loom as warp threads. Weaving was followed by bleaching, dyeing, washing, and stretching.[41]

In the eleventh century, most of Europe's silk was still

imported from Islamic lands or Asia, despite a sizable silk industry in Byzantium. The Church was a large customer, favoring silk as a wrapping for the relics of its saints and ignoring occasional Arabic inscriptions: on St. Cuthbert's shroud in Durham Cathedral, "There is no God but Allah"; on that of St. Josse of Brittany, "Glory and rising fortune to the commander Abu Manu Haidr; may Allah lengthen his life."[42] By the twelfth century, silk manufacture too began to move into Italy. Venetians had already established their own silk-weaving shops in Constantinople, where they enjoyed a privileged position; the next step was to introduce silk craftsmen into Italy. Chronicler Otto of Freising tells us that in 1147 Norman king Roger II of Sicily brought weavers of silk from several Greek cities to Palermo. A third center, Lucca, had specialized in the silk trade in the eleventh century and succeeded in founding its own industry in the twelfth.[43]

Other industries besides textiles tended to concentrate by city or region: fine glass on the island of Murano (Venice), pots and pans in Flemish Dinant, arms and armor in Nuremberg and Milan. By the twelfth century, Milan was "a kind of general arsenal town" (Robert Reynolds) producing in large quantities armor, weapons, horseshoes, nails, and crossbow bolts.[44]

Most cities were less specialized, however. Gawain, hero of Chrétien de Troyes's Arthurian romance, Le Conte del Graal, gazes at a town with its "many fine people" and notes their vocations: "This man is making helmets, this one mailed coats; another makes saddles, and another shields. One man manufactures bridles, another spurs. Some polish sword blades, others full cloth, and some are dyers. Some prick the fabrics [raise the nap] and others clip [shear] them, and these [men] here are melting gold and silver. They make rich and lovely pieces: cups, drinking vessels, and bowls, and jewels worked in with enamels; also rings, belts, and pins."[45] Knightly Gawain's eye is attracted to the crafts of war and luxury, but he might have noted some hundred others in a prosperous twelfth-century town, including cordwainers (shoemakers), cobblers (shoe repairers), tailors, carpenters, leather workers, coopers, masons, butchers, brewers, and in smaller numbers hatters, harness makers, cartwrights, purse makers,.and glove makers. The goldsmiths that he mentions worked not only gold and silver but tin, brass, and copper, hammering, sawing, smoothing, polishing to fashion rosaries, cups, ornaments, and the gold leaf that adorned illuminated manuscripts. Tanners scraped hides, rubbed them with dung to soften, and either soaked them in tannin baths with oak bark, oak galls, or acacia pods or transformed them into white leather by tawing—treating with alum and salt. The techniques were known to the ancients and not substantially improved until the nineteenth century.[46]

Thus, although only cloth making achieved the status of a large-scale commercial industry, handicraft production was a flourishing institution by the twelfth century, and despite the prejudice of the landed aristocracy against those who worked with their hands to earn money, it enjoyed a considerable measure of respect. This respect reflected the distinctly medieval character of the work, combining concern for quality with pride in individual craftsmanship. Every product was in a sense a work of art. Tools, often made by the craftsmen themselves, were precious and were passed down from father to son or bequeathed in wills to favored colleagues.

In the twelfth century, the guilds of the merchants were joined by guilds of craftsmen. These craft guilds, with their hierarchy of masters, journeymen, and apprentices, had two principal functions: mutual aid and control of production— quality, price, working hours, and wages. Each guild monopolized its craft, but at first membership was relatively easy to attain, and the controls exercised were moderate. The charge that the craft guilds resisted technical progress applied only to a later age and even then derived mainly from the guilds' concern for handicraft quality. "In so far as [the craft guilds] succeeded in regulating growth without stopping it," writes Robert Lopez, "they spared their humbler members the extreme sufferings that were inflicted on the slave gangs of antiquity and the factory hands of the early Industrial Revolution."[47] New machines and processes, designed to speed production and save labor, inevitably provoked the craftsman's suspicion that quality would suffer along with his livelihood. Nevertheless, he was by no means always blind to the value of technical innovation. John Harvey credits the medieval building trades with "the introduction of new inventions, new processes, and new aesthetic ideas after about 1100."[48]

Among the craft innovations in the central Middle Ages was the drawplate, a device that aided blacksmiths in fabricating wire for chain mail, until then laboriously hammered out at the forge. The piece of heated iron was drawn through successively smaller holes, a process that remained only to be mechanized. Another innovation took place as soap became an article of commerce, by the substitution of olive oil for animal fats. As a result the industry migrated south from its early center in Scandinavia to Marseilles, Venice, and Castile.[49]

The only significant modification ever made in the art of brewing came with the introduction of hops, as early as the tenth century in some places, as late as the fourteenth in others. Hops imparted the faintly bitter flavor that gradually won preference over the fruity taste of the older brew and also served as a preservative. Among medieval crafts, brewing had two distinctive characteristics. Its universality in beer-drinking regions, second only to spinning and weaving, made it a village as well as a town vocation; and in both town and village a high proportion of its practitioners were women.[50]

Apart from brewing, the craft world was dominated by males, but in the family production unit that prevailed, wives shared work with their husbands and widows often succeeded them. Some married women worked at trades of their own; single women occasionally worked at crafts as *femmes soles* and enjoyed a status of recognized equality.[51]

Two crafts that came to be practiced in villages and on rural estates nearly everywhere were those of carpenter and smith. The village carpenter was probably as old a feature as the village itself, fashioning all kinds of implements as well as components of buildings, bridges, and wagons by splitting, trimming, sawing, hewing, and shaving the logs he cut from the neighboring forest. The smith, long primarily armorer for the castle elite, shifted to the village only as demand for his services to agriculture grew. A study in Picardy showed no trace of a village smith before 1125 but counted thirty by 1180.[52] The combination of the two crafts was indispensable for making and repairing agricultural implements—plows, spades, mattocks, hoes, axes, billhooks, sickles.

Commonly the two trades collaborated, since most of their products combined wood and iron. The carpenter used nails handmade by the smith for his fastenings, usually boring a hole first with an awl.[53] Saws, hammers, axes, adzes, drills, and knives required precision work by both smith and carpenter. Heating the workpiece at his forge, the smith hammered it on an anvil mounted on a stout trunk of wood, using sledgehammer (often wielded by an assistant or apprentice), lighter hammers, tongs, pritchel (punch), bellows, and file, besides a stone trough for quenching and a grindstone for sharpening and smoothing. Chisels were either "cold" or "hot"—designed to work pieces cold or red-hot—the cold chisel needing a harder edge. Tempering by repeated heating and quenching gave hardness and ductility but was time-consuming; only tools used on stone or metal were given a hard temper. Precision was strictly a matter of hand-eye coordination. A file was made by neatly striking a succession of closely spaced blows on a flat piece of heated iron with a sharp hammer. Saws also demanded patience and skill to produce sharp and even teeth along with overall hardness and flexibility. The saw was a carpenter's tool; the peasant's cutting tool, the ax, was easier to make.[54]

As building construction and commerce increased, the smith acquired important new customers: masons (mallets, picks, wedges, chisels, cramps, stays, tie-rods, and dowels); carters and wagoners (iron parts); millers (iron components of hydraulic machinery); and shipbuilders (nails and fittings).

Besides town and village, the great lay estates retained their own craftsmen—armorers, fletchers (arrow makers), smiths—while the monasteries maintained regular workshops where master glaziers and enamelers as well as goldsmiths exercised their arts. A description of one such monastic workshop is a feature of a unique treatise, Europe's first technical manual, *De diversis artibus* (On diverse arts), composed in the first half of the twelfth century by Theophilus Presbyter (Theophilus the Priest), believed to be a pseudonym for a Benedictine monk named Roger of Helmarshausen, known for his metalwork artistry.[55]

Unlike the numerous authors of encyclopedic, alchemical, and even practical manuals of the day, Theophilus was distinguished by his reliance on knowledge gained from firsthand experience. Addressing himself to "all who wish to avoid and subdue sloth of mind and wandering of the spirit by useful occupation of the hands and delightful contemplation of new things," he divides his manual into three books. In the first, "The Art of the Painter," he gives minute directions for making a variety of pigments, along with varnish, gold leaf, and ink. His recommendation of "Byzantine parchment" is probably the first documentary reference in the West to paper. Book Two,

Armorer at work: a spinoff from the blacksmith's craft. [Trinity College, Cambridge, Ms. 0.9.34, f. 24.]

"The Art of the Worker in Glass," describes a scale of operations that implies a work force of at least twelve and cites many
specialized tools, including the blowpipe. Instructions follow on
glass furnaces, proper mixtures of ashes and sand, manufacture
of glass sheets, flasks, goblets, and windows, and repair of glass
vessels.

But Book Three, "The Art of the Metalworker," is clearly
nearest the author's heart. Devoting twice as much space to
metalwork as to the first two subjects combined, Theophilus
begins with a detailed description of the windowed workshop,
partitioned into rooms for casting and working base metals,
gold, and silver, each with its own specially designed work-
benches and forges. He tells how to fashion bellows from ram
skins and describes anvils, hammers, tongs, pliers, drawplates for
wire, files, punches, chisels, and other implements. Theophilus
gives "the earliest good description of the very ancient art of
cupellation," the oxidation of lead to separate out gold and silver.[56] Only rarely does he wax fanciful; in his recommendation
of goat's urine or that of "a small red-haired boy" for quenching
hot metal, he follows a "long and honorable" tradition of
medieval metallurgists (Nadine George).[57]

All the punching, sinking, chasing, engraving, and repoussé
that Theophilus so affectionately describes depended for its
success on the skill and experience of the operator, expressed
through his array of hand tools. Despite the implied use of a
crank for rotary motion, nowhere does any large machine
intrude, with the possible exception of a sort of lathe to shape
molds in (bronze) bell founding. The sole reference to mechanical power is to the wind employed by the pipe organ.

In the twelfth century, much of metal production, from digging
the ore to hammering out the finished product, remained a
matter of laborious manual effort. Surface deposits of iron ore
were no longer sufficient, and pits, trenches, and tunnels were
driven into the earth, especially in central Europe. Slaves had
disappeared from mine and forge, but large-scale capitalist organization of work, with its scope for power technology, lay in the

future. The form that organization took in the "customs of miners" of central Europe and north Italy was characteristically medieval. First written down in 1185 in Trento, the "customs" followed the pattern of manorial agriculture: a share of each seam of ore was assigned to each family of miners, while decisions as to working methods, hours of work, and division of profits were made collectively by representatives of the miners and of the prince or landowner who was their lord.[58] Precious-metal miners were paid in ore, which they sold to waiting gold- and silversmiths at the end of each day.[59]

A Bible in Glass and Stone

Much of the craft described by Theophilus Presbyter reflects the contemporary enrichment of churches and monasteries by window glass, painting, and metalwork. Lords and merchants contributed gifts and bequests to benefit their souls, guilds donated windows showing the guildsman at work at his craft,[*] and pilgrims made contributions in honor of saints' relics. The wave of philanthropy merged with a series of architectural inventions in the flowering of a new style of church architecture. Whatever their religious meaning, the towering new cathedrals signaled affluence. Of the eighty new-style cathedrals built in France from 1150 to 1280, a large proportion rose in the prosperous cloth towns of the North.[60]

The new architecture had its earliest beginning in the tenth century, in Burgundy. There the great abbey of Cluny, founded in 920 to spearhead reform of the old Benedictine Order, rebuilt its vast mother church (c. 980), taking note of two recent tendencies in Western Christianity: the custom of daily mass and the cult of the saints. At the eastern end of the abbey church, an ambulatory, or semicircular passage, was added, with chapels radiating off it for celebration of masses and preservation of saints' relics. Elsewhere, new churches provided extra chapels by extending their side aisles past the transept.[61]

[*]Not, however, in the case of the guild of prostitutes in Paris, who made an unobtrusive gift to Notre Dame.

At the same time, better ways of supporting the massive Roman barrel vault were sought. From such Roman monuments as the Baths of Caracalla, builders revived the groin vault, formed by making two semicircular arches intersect at right angles, so that the weight of the vault rested on four massive corner columns instead of on two walls. An alternative was to rest the vault on transverse arches, that is, a row of arches running along either side of the nave. The result was the eleventh-century Romanesque triple vault, in which two narrow side-aisle vaults flanked the broader and higher nave vault, forming a mutual support system. As builders acquired fuller mastery of their technique, Romanesque vaults widened and rose.[62] A second rebuilding of the church at Cluny in 1088 (Cluny III to modern architectural historians) created a nave 40 feet wide and 98 feet high, while a new church built at Speyer, on the Rhine, measured 45 feet in width and 107 in height.[63]

Back in the sixth century, Pope Gregory the Great had made a plea for depicting scriptural scenes on church walls for the benefit of the unlettered faithful. A synod at Arras in 1025 reiterated the recommendation, for "this enables illiterate people to learn what books cannot teach them." But wall paintings in barrel-vaulted churches were hardly discernible in the dim light. When the nave was raised above the side aisles, a row of windows could be added, but too high to contribute much illumination.[64]

Thus church builders were pressed to create larger churches, with more complex floor plans and better light. The economic expansion of the eleventh and twelfth centuries underwrote a period of tireless experiment and rewarding discovery in masonry construction. The architectural style that emerged grew out of need, inspiration, and the accumulation of technical resources. Foremost in importance of the last were three engineering devices: the pointed (ogival) arch, the rib or cross-rib vault that derived from it, and the flying buttress that lent it external support. All three were developed separately in Romanesque buildings before being united to form what came to be called, much later and quite inappropriately, "Gothic" architecture.

The pointed arch is now believed to have originated in India and to have migrated westward to reach Italy by the eleventh century. In 1071 a porch (narthex) of the new abbey church at

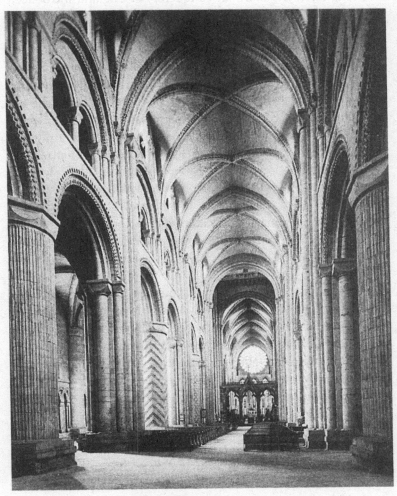

Nave of Durham Cathedral, combining Romanesque side arches in the nave and the triforium gallery with a Gothic cross-rib vault. [Royal Commission on the Historical Monuments of England.]

Monte Cassino, birthplace of the Benedictine Order, was given pointed arches and rib vaults. The result was a triumph. By distributing the weight of the vault to a skeletal structure of vertical columns, it converted walls from supporting elements into mere panels and opened the possibility of large window spaces. Not long after, the Norman conquerors of England introduced rib vaults in the choir of the cathedral they had begun at Durham in 1093, though keeping the arches round; later, in the nave, they made them pointed.

Meanwhile, Cluny's Abbot Hugh visited Monte Cassino before embarking on Cluny III, a structure still Romanesque but incorporating several pointed arches. These caught the eye of visiting Abbot Suger, of the royal abbey of St. Denis, where the kings of France were buried. As he recorded in a memoir that provides rare documentary light on the development of Gothic construction, Suger was looking for ideas for the rebuilding of his own abbey. Beginning work in 1135, he discarded the Romanesque style completely, signaling the full arrival of the dramatic new architecture: pointed arches, rib vaulting, soaring vertical lines, and enormous window spaces filled with stained glass, including a huge rose window in the west front.[65]

Suddenly Pope Gregory's wish found a dazzling fulfillment. "A blaze of glass windows" was a twelfth-century chronicler's description of Canterbury Cathedral following its rebuilding in the new style.[66] A few conservative churchmen even found the new brilliance too glaring, but the public enthusiastically applauded the stained glass, manufactured in the forest, where fuel and raw materials were available and where the smoke provoked no protest.

Twelfth-century glassmaking followed long-established methods, although sodium ash from Europe's hardwood forests had replaced potassium ash as principal ingredient. Two parts of wood ash (beechwood preferred) mixed with one part sand were heated in a furnace to a moderate temperature that produced a reaction between the sodium carbonate in the ash and the silica in the sand, a process called "fritting." The "frit" was melted by

Glassblowing, from a twelfth-century version of a work of Hrabanus Maurus. [Division of Rare and Manuscript Collections, Carl A. Kroch Library, Cornell University.]

a higher temperature and blown with a long iron tube. To produce window glass, the resulting bubble was shaped into a cylinder, the ends cut off, and the cylinder split down the long axis and flattened to a sheet. An eleventh-century manuscript illustration shows a furnace with three tiers, the lowest for fritting, the middle for melting, and the top for annealing, the process of controlled cooling.[67]

Like metallurgy, glass manufacture was a rule-of-thumb process, relying not on knowledge of chemistry but on trial and error. At first colors were obtained by varying the proportions of raw materials and by changing the melting time—such at least is the conclusion drawn from the description of Theophilus Presbyter. "If [the melt] happens to turn a tawny flesh-like color," Theophilus wrote, "heat for two hours and it will become reddish-purple and exquisite." Later the addition of metallic oxides made colors truer and more easily controlled: cobalt for blue, manganese for purple, copper for red, iron for yellow. More difficult to achieve, and in fact for a long time unobtainable, was clear glass. By cutting up the sheets of colored glass, glaziers

An early-fifteenth-century Bohemian forest glasshouse. The main furnace is being stocked by the small boy on the right, while annealing furnace is on the left. Melted glass is collected (second from right) and blown (center). Behind the furnace, the master glassmaker inspects a vessel. In the background, sand is being quarried and brought to the furnace. [British Library, Ms. Add. 24189, f. 16.]

could create designs, which almost at once became pictorial. Out of accident and need, art was born.[68]

Glassmakers clustered in areas with suitable forests—such as Normandy, Burgundy, Lorraine, Germany, and Flanders—but traveled freely to meet demand for their services. Most of the craftsmen who made the windows of medieval English churches and cathedrals were imported from the Continent, like Laurence Vitrearius (Glassmaker), a Norman who created the windows in the east end of Westminster Abbey.[69]

The third Gothic element, the flying buttress, was originally devised to shore up Romanesque walls threatening to collapse outward. Conventional buttresses—extra thicknesses added to walls at weak points—had been employed since Roman times but were unusable in a church with side aisles. The flying buttress proved not only an effective but an elegant solution. Leaping airily over the low side aisle, it caught the nave vault at the critical point where the outward-thrusting arch came to rest on the pier. The two thrusts neatly counterbalanced each other.[70] That the flying buttress was also an aesthetic triumph was accidental but quickly recognized. Incorporated in the original design of Notre Dame de Paris, it proved a crowning touch. Stone caps, added to press the buttress down firmly, came as another aesthetically happy accident. Builders soon realized that interior piers could now be made slimmer. Thus innovations based on engineering imperatives rather than religious or artistic considerations provided the basis for the impressive new combination of religious expression and architecture.

From St. Denis, Durham, and Notre Dame de Paris, the new style spread across northwest Europe. In its final form, the Gothic church rose from a ground plan in the form of a cross, with nave (space for the congregation) and choir (space for the

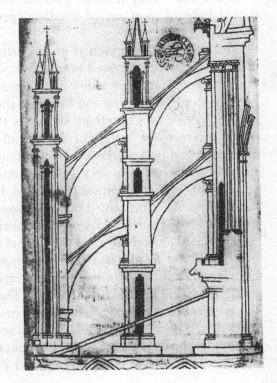

Double flying buttress at Rheims Cathedral, as sketched by thirteenth-century master mason Villard de Honnecourt. [From The Notebook of Villard de Honnecourt, ed. by Theodore Bowie, Indiana University Press.]

clergy) separated by a transept, and with an apse at the eastern end, usually consisting of an ambulatory and a number of chapels.

Abbot Suger's memoir describes some of the procedures in the rebuilding of his church of St. Denis. The discovery of a new quarry "through a gift of God" yielded "very strong stone" for the construction. Equally providential was the arrival of "a skillful crowd of masons, stonecutters, sculptors, and other workmen." The abbot himself and his carpenters scoured the abbey's forests for suitable timbers for the roof and regarded it as a miracle when they found the twelve great beams they needed. "Moreover, it was cunningly provided" that the old nave and the old side aisles should be correctly aligned with the new ones "by means of geometrical and arithmetical instruments," while the circular string of chapels in the ambulatory would make the whole church shine "with the wonderful and uninterrupted light of most sacred windows, pervading the interior beauty."[71]

One detailed description of the construction of a twelfth-century church has survived from an account of the rebuilding of Canterbury Cathedral following a disastrous fire in 1174.[72] In his "Tract on the Burning and Repair of the Church of Canterbury," Gervase of Canterbury describes the arrival "amongst other workmen" of "a certain William of Sens, a man active and ready and as a workman most skillful both in wood and stone," of "lively genius and good reputation." Other "French and English artificers," already consulted, had given contradictory advice. William studied the precarious state of the surviving parts of the church and determined that the entire structure must be razed and rebuilt from the ground up. He waited, however, until the monks had recovered from the shock of the holocaust before he gave his report, meanwhile "preparing all things that were needful for the work." The monks, daunted by the prospect of a complete rebuilding, nevertheless consented "patiently, if not willingly," and William set to work, importing stone from the Continent—probably from Caen, in Normandy, a prime source of building stone—and preparing "molds for

*Twelfth-century building construction is shown in
representation of the Biblical Tower of Babel. No
lifting machinery is in evidence, masons carry hods,
and primitive scaffolding is made from tree trunks.
[Division of Rare and Manuscript Collections,
Carl A. Kroch Library, Cornell University.]*

shaping the stones" for the assembled sculptors. Meanwhile the
choir of the old church was razed.

Gervase describes the course of the reconstruction year by
year: in the second year, 1175, the pillars and arches of the
choir and the vaults over the side aisles were erected; in the
following year, the main pillars at the meeting of choir and
crossing and the gallery (triforium) above; the next year ten
pillars of the nave and their corresponding arches and triforium.

One day in the fifth year of construction, William had
climbed the timber falsework to the height of some fifty feet to
direct the placing of "machines for the turning of the great

William of Sens's choir, Canterbury Cathedral. [Royal Commission on the Historical Monuments of England.]

vault," when the scaffold gave way under his feet and he plunged to the floor amid a shower of stones and timber. Bruised and shaken, he took to his bed. The doctor was summoned, but the patient's condition did not improve. With winter approaching and the upper vault yet to be finished, William chose "a certain ingenious and industrious monk" as his assistant and directed the work through him, an arrangement that caused some "envy and malice" among the other workers. When recovery continued to elude William, he resigned his post and returned to France. He was succeeded by another William, called "the Englishman," "small in body, but in workmanship of many kinds acute and honest." Under him the work proceeded for five more years. First the choir was finished and the altars erected so that Easter service might be performed there immediately. Later the exterior walls were built, a new crypt was completed, and the tower raised. "In the ninth year [1184] no work was done for want of funds." In the tenth the cathedral was at last finished, the tower covered over, and the roofs protected with lead.

Gervase summed up the improvements of the new Gothic-style church over the old. The whole structure was much higher; the new pillars were taller; their capitals were sculptured where the old ones were plain. Marble columns were added, and the vaults were arch ribbed, with keystones. For the wall that had separated the choir from the transept, hiding it from the nave, were substituted pillars and a vault that left it open to view. Instead of a wooden ceiling, the church was covered with "a vault beautifully constructed of stone and light tufa [porous limestone]," and a second triforium gallery was added in the choir and two more in the side aisles of the nave. The effect was one of height, space, decoration, and light.[72]

St. Bernard, whose Cistercian Order played a leading role in diffusing the Gothic style, thought the soaring height and sumptuous ornamentation inappropriate for monastic buildings, but even St. Bernard could see the value for the laity of the Bible in glass and stone.[73]

Castle, Trebuchet, and Crossbow

The period that witnessed the evolution of Gothic architecture saw equally impressive advances in military engineering. Development of the masonry castle that largely supplanted the old timber motte-and-bailey was not stimulated by increased peril from outside. On the contrary, it coincided with the conversion of the Vikings into peaceful settlers and traders, and with Europe's taking the offensive against Islam in the Mediterranean, in Spain, and with the First Crusade (1095–1099), in the Levant.

What may have been the first stone "keep" in Europe was built by Fulk Nerra, count of Anjou (c. 970–1040) at Langeais, on the Loire near Tours, about the year 1000. In the eleventh century many more such keeps appeared, lofty, formidable, threatening, and like the cathedrals enduring monuments of the age. Square or rectangular in plan, the keep was usually three stories high, built of stone blocks in courses enclosing a rubble core, the floors timber, the windows few and narrow.

Rochester Castle, rectangular keep with a parapet 113 feet high, built in 1130. The round tower at left replaced a square tower undermined by King John in the siege of 1215.

Another type of masonry castle, the "shell keep," appeared a little later: a circular stone wall, sometimes crowning a mound, within which living quarters were built in timber or stone.[74]

The rectangular keep and shell keep reflected the new feudal order, in which Europe was ruled by local lords whose castle-domiciles commanded the surrounding countryside. In England at the accession of Henry II in 1154, the king held 49 castles, the barons of the realm 225. The gradual shift of power toward the king may be read in the figures of sixty years later, at the time of Magna Carta: 93 royal castles, 179 baronial.[75]

These fortresses were not, however, the last word in castle architecture. The advance to the final stage resulted in no small measure from the First Crusade. By leaving a few thousand knights and men-at-arms marooned in Syria and Palestine amid a sea of hostile Saracens, the Crusade placed a tremendous burden on defensive fortifications. Enriching their own European experience with ideas borrowed from the enemy—entrance traps, concentric walls, inner keeps[76]—the Crusading orders of the Templars and Hospitalers lined the frontiers of Syria and Palestine with castles that were the engineering marvels of the age, sometimes built from scratch, sometimes by

Shell keep at Gisors, Normandy, early twelfth century, built on an artificial mound forty-five feet high. Four-story octagonal tower was added later by English king Henry II.

enlarging, repairing, and elaborating captured Arab fortresses. Where earlier castle builders in Europe had paid scant attention to terrain, the Crusader castles were carefully sited on high ground, their few approaches cleared of cover and broken by a deep ditch (the moat). The corners of the rectangular keep were vulnerable to sappers digging under them and offered "dead ground" for attackers armed with battering rams to shelter out of reach of the garrison's missiles; the Crusading castles, copying Byzantine and Saracen models, gave their corners round or multisided towers that effectively resisted both threats.[77]

The main line of defense of the new castle was its high curtain wall, whose parapet was protected by crenellated battlements and provided with machicolations, another borrowing from the Arab enemy—openings in the floor through which missiles and boiling liquids could be dropped. Firing windows, called "arrow loops" or *meurtrières* (murderesses) were flared to the inside, giving the defending archer room to move laterally, with a wide field of fire, while presenting to the besieger only a narrow exterior slit as a target.

Entry to the castle was through a gatehouse, whose

Crenellated curtain wall, with machicolations—openings through which missiles could be dropped. Fougères, Brittany.

A firing window or arrow loop flared to the inside (with late modification for guns). Falaise, Normandy.

portcullis, oak plated and shod with iron, was raised vertically by a pulley operated from an upper chamber in the gatehouse. Spanning the moat was a bridge with a draw section operated by counterweight. A device unknown to the Romans, the drawbridge was forgotten after the Middle Ages until revived by nineteenth-century bridge engineers for its modern purpose of spanning busy waterways. Sometimes the entry was through a skewed passage where infiltrators might find themselves trapped under fire from archers above them. If the outer wall was breached, the defenders could still fall back to the inner keep, with its own well-planned fields of fire.

The most famous of all the twelfth-century castles in Europe and Syria was the Krak des Chevaliers (Citadel of the Knights). Built by the Hospitalers in the first half of the twelfth century on the spur of a mountain, its massive walls formed two concentric rings dominated by great towers and separated by a wide moat. In addition to *meurtrières*, crenellations, and machicola-

The Krak des Chevaliers (Citadel of the Knights), Crusader castle in the Holy Land. [From Kenneth M. Setton, ed., A History of the Crusades, University of Wisconsin Press.]

tions, the castle's walls bore a less ferocious piece of technology, a windmill to grind the garrison's grain.[78] Besieged at least a dozen times, the Krak resisted every assault and continued to "stick like a bone in the very throat" of the Saracens, in the words of a Muslim writer,[79] until in 1271, one of the last European strongholds in Asia Minor, it was captured by a ruse. The Muslim general Baibars tricked the garrison with a forged order and afterward chivalrously provided them with safe-conduct to the coast.[80]

The living quarters of twelfth-century castles featured a device that was immediately and widely copied: the wall fireplace. A marked improvement in efficiency over the old central hearth, the fireplace provided heat both directly and by radiation from its own stones and from the wall opposite. Late in the century, a projecting hood was added to better control the smoke, and the sides of the hearth were splayed to increase radiation. Whether in a castle or in a house, the fireplace was immense by modern standards, designed to accommodate large, long-burning logs.[81]

Despite their size and complexity, castles were often built

Wall fireplace in great hall of the early-twelfth-century square keep at Vire, Normandy.

with impressive speed if the need was urgent and the financial resources were available. Richard Lionheart's Château Gaillard on the Seine was essentially built in a single year.[82]

To attack or defend a castle, missile weapons were indispensable, and the period that initiated the building of castles also saw the introduction of a new form of artillery. A version of the trebuchet employing human traction to power the firing beam was invented in China as early as the first century A.D. and possibly passed via the Turks to the Arabs and Byzantines.[83] Whether Parisians employed it against the besieging Vikings in the ninth century is a minor scholarly controversy. In the period between 1180 and 1220, the tre-

Loading a trebuchet, top left; adjustable counterweights are hidden behind melee at center. [The Pierpont Morgan Library, Maciejowski Bible, M. 638, f. 23r. (detail)]

buchet was given a decisive improvement in Europe by the substitution of a huge counterweight for the manpower brigade. The machine's arm was mounted asymmetrically on a fulcrum, with the short end, pointing toward the target, given the counterweight. The long end, wound back by a winch, was released by a blow from a mallet. The missile was carried in a sling, attached by long lines and lying at rest in a trough under the machine; when triggered, the beam sprang upward through an arc, gaining acceleration before the missile was picked up. Consequently, high "muzzle velocity" could be achieved, especially if the missile was released at near the optimum angle of 45 degrees.

A vast amount of experimentation, of which no record survives, must have gone into the development of the counterweight trebuchet. The counterweight was in the form of a hopper filled with earth or stone, specified in one source as "nine feet across and twelve feet deep."[84] Varying the weight controlled the range. Modern experiments comparing the trebuchet with its ancient forebears have shown the medieval weapon capable of hurling a much larger projectile, a mass of

100 to 150 kilograms (220 to 330 pounds) a distance of 150 meters (160 yards).[85] The older catapults have shown greater range, up to 225 meters (245 yards), but only with a much lighter missile, one of 20 to 30 kilograms (44 to 66 pounds). To batter down heavy masonry walls, weight of projectile was much the more important consideration. The value attached to the trebuchet is reflected in the custom of giving each engine a name, such as those recorded for Edward I's siege of Stirling in 1304: Vicar, Parson, War-Wolf, Gloucester, Belfry, Tout-le-monde.[86]

A different version of the same machine, the "mangonel," appearing at the same time, had a lower "angle of departure," giving its missile a flatter parabola. The effectiveness of trebuchet and mangonel is indicated by the absence after the tenth century of references to the old-fashioned torsion engines.[87]

An even more important missile weapon made its appearance in Italy in the eleventh century. The crossbow had been known to the Romans (and the Chinese) centuries earlier but had never been very effective in its ancient form and had disappeared from warfare. Its basic principle was that a bow set transversely on a stock, or crosspiece, could be bent farther and so develop greater muzzle velocity than a handbow. In its medieval reincarnation, the weapon was cocked by resting the bow on the ground with the stock upright; the archer placed his foot in a stirrup on the stock, stooped to catch the bowstring on a hook in his belt, and by straightening up bent the bow, using the strength of his whole upper body instead of merely that of his arm. The bowstring was brought back to a locking device on a groove in the stock and fitted to the bolt, a short, very thick arrow armed with a heavy iron tip.

The crossbow was more expensive than the ordinary bow and took longer to reload, but in castle defense both disadvantages were minimized. The archer could reload in safety, while in the economics of siege warfare, in which a garrison of a hundred was expected to hold out against thousands of attackers, high cost of individual equipment was justified.

An important improvement in the medieval crossbow may have been learned during the Crusades from the Saracen enemy (strangely, the Crusaders' Byzantine allies in 1097 knew nothing of the weapon, which princess-chronicler Anna Comnena described as "a Frankish novelty" and "a truly diabolical machine").[88] Wood as bow material was strong in compression but weak in tension, and the outer part of a bent bow was under tension. The tension created by the strength of a man's arm was easily resisted by a strong wood, such as yew. But the tension created by the crossbow's bending could cause "slithers" to splinter away. A strip of animal sinew, usually the large ligament that runs along the spine of most mammals, was incorporated as a reinforcing layer on top of the wood, absorbing some of the tension. Such a "composite" bow was made even stronger by a layer of animal horn on the underside. The power of the new weapon alarmed the Church, which in the Lateran Council of 1139 anathematized its use (and for good measure that of bows in general) against Christians. "Naturally, this prohibition was very unevenly observed" (Philippe Contamine).[89]

The Brothers of the Bridge

Crusaders, pilgrims, and merchants were perpetually on the road, but travel in the age of the Crusades was not for the faint-hearted. A tenth-century wayfarer described his journey from Constantinople to Lepanto, Greece, a distance of some five hundred miles, in these terms: "On mule-back, on foot, on horseback, fasting, suffering from thirst, sighing, weeping, lamenting, I arrived after forty-nine days."[90] By that time it was evident that the old Roman road network was out of date as well as out of repair. Its steep grades hindered merchants' carts and pack animals, and its crumbling surfaces challenged even unencumbered travelers. Many of the routes, designed to serve Roman garrisons, no longer went where people wanted to go, while the new and growing traffic of medieval pilgrims needed roads to places the pagans had never sought to visit—Canterbury, Compostela, Roc Amadour.

As medieval commercial expansion altered the ratio of traffic, wheeled vehicles for the first time demanded a substantial share of the road. Certain kinds of merchandise suffered from rough surfaces. The tight-fitting wooden barrels that had replaced amphorae allowed wine to age properly but were likely to burst under severe jolts.[91] How much road construction and maintenance was practiced in the central Middle Ages is unknown, but road builders did develop a surface of cobbles set in a thick cushion of sand that was better adapted to the northern climate than the rigid Roman pavement. As central and regional governments strengthened, roads under royal, imperial, or other seigneurial protection were most likely to receive attention. In England, royal roads were supposed to be maintained at a width sufficient for two wagons to pass, or for two oxherds to make their goads touch, or for sixteen armed knights to ride side by side.[92] The only road built from scratch under the Norman kings of the eleventh and twelfth centuries was the one cut by Henry I in 1102 for the passage of his army over Wenlock Edge.[93]

By the twelfth century, the horse collar and harness had made the horse much the preferred traction for cart and wagon. Maneuverability was improved by several new or revived features: substitution of a pair of shafts for the old single draft pole; the whippletree, a transverse bar pivoted in the middle and positioned in front of the wagon (illustrated drawing a plow in the Bayeux Tapestry);[94] and the pivoted front axle. Nailed horseshoes, common by the eleventh century, were mass-produced by the twelfth.[95] The spread of stirrups and saddles helped make riding easier and more popular.

In bridge construction, a tenth-century "nadir" (Marjorie Nice Boyer) was followed by a resurgence in the eleventh century.[96] The eclipse of the Carolingian monarchy and the inheritance by local lords of the wardenship of river crossings opened up a new chapter in bridge building. Increased volume made toll collection an important source of revenue for many lords, while drawing attention to the inadequacy of ferries, where wagons, pack trains, droves of animals, and troops of the pious queued up to board the skiff or raft.

The Church intervened. Bishops began granting indulgences for bridge construction and repair, and the monastic orders took to collecting funds as well as maintaining hospices at crossings.[97] Bridges also profited from inclusion in legacies, along with churches, hospitals, and "other pious and poor places."[98]

In southeastern France in the twelfth century, a new, highly specialized order was founded by an ex-shepherd who became St. Bénézet, also known as Little Benedict the Bridgebuilder. The Frères Pontifes, or Brothers of the Bridge, built the Pont d'Avignon, whose twenty arches spanning the Rhône (and the island of Barthelasse) were of a new design, credited to Bénézet himself: elliptical, with the long axis vertical. The tall arches required less support during construction than the old Roman semicircular arches, permitting narrower piers in the stream. At the same time they provided more room for the rising waters of the notorious Rhône floods. The decrease in constriction reduced scour around the piling that underpinned the piers, the main threat to the stability of stone-arch bridges.[99]

Remaining four arches of the Pont d'Avignon, over the Rhône, built by the Brothers of the Bridge in the twelfth century. [French Government Tourist Office.]

Pont-St.-Esprit, another Brothers of the Bridge construction, with flattened segmental arches.

Near the Avignon end of his bridge, St. Bénézet built a combination of chapel and toll station, mingling piety with practicality in a graceful symbol. The Brothers of the Bridge raised funds and oversaw construction of at least two other large stone-arch bridges, at Lyons and Pont-St.-Esprit. The St. Esprit span introduced a design new to Europe, though long known in China: the flattened segmental arch (based on an arc smaller than a semicircle). The opposite of St. Bénézet's tall arch, it was difficult to construct and had been avoided by the conservative Romans, but once in place it derived superior stability from the smaller number of piers in the river and consequent reduction in scour.[100] The St. Esprit arches, however, were only slightly flatter than semicircular, and the extent to which the bridge's builders understood the advantage is hard to say. Until the fourteenth century, nearly all medieval bridges continued to employ the semicircular arch.

In the cities medieval bridges increasingly took on the function, unknown to the bridges of Rome, of supporting houses and shops, for which they offered convenience in both water supply and sewage disposal. London Bridge, begun in 1176 by a local chaplain named Peter of Colechurch and partly financed by a thousand-mark gift from the papal legate, became the picturesque heart of the city, its roadway loaded with a double row of structures of varying sizes and purposes. A less sophisticated design than the Pont d'Avignon, London Bridge was supported by nineteen semicircular stone arches, no two quite alike, built summer after summer as money was available, until completion in 1209. Its massive piers so constricted the tidal Thames that piloting a boat through at high water became a sporting proposition known as "shooting the bridge." A small draw span, included as a defensive measure, was too narrow to allow most boats to pass. London boatmen adopted the expedient, quickly copied elsewhere, of removable masts for squeezing under the bridge. But the occasional use of the draw span to pass narrow-hulled, fixed-mast craft may have been the first application anywhere of the drawbridge to navigation.[101]

Construction techniques had changed little from Roman times. The season for work ran generally from mid-June to September-October. The principal machinery used was a ram, a heavy weight raised by windlass or pulley that drove piles, three to four yards long and sharpened at the ends, into the riverbed.

The bridges of Paris: on the Grand Pont, left, a money changer awaits customers, a goldsmith hammers, and a rag-picker carries basket and pick; on the Petit Pont, right, a doctor gives a prescription in an apothecary shop; on the river below, a fisherman casts his line. [Bibliothèque Nationale, Ms. fr. 2092, f. 35v.]

The superstructure was often a mixture of stone and wood, the latter chosen with care, with preference for chestnut, which resists rotting and is strong in compression. A roadway might be supported by timber arches sprung from stone piers or by a combination of stone arches and timber trestle.[102]

Medieval builders introduced an improvement in the starlings, or cutwaters, the footings from which the piers rose. Roman starlings were pointed at the upstream end to breast the current but made square downstream. St. Bénézet may have been the first to point his downstream starlings, an innovation that reduced eddying and so helped combat scour.[103]

The Pont d'Avignon stood intact until the seventeenth century, London Bridge until the nineteenth, and the Pont-St.-Esprit and several others still stand today. Despite numerous failures from flood, scour, fire, ice floes, and enemy action, medieval bridges may be judged eminently successful solutions to a set of engineering problems under the handicap of existing conditions.

In the teeming walled towns, the bridge was typically the center of activity. A person could be born on a bridge, grow up there, work in his establishment and live upstairs, go to church, shop at the butcher's and baker's, and retire to a hospital for the aged, all without leaving the bridge. In Paris he might enjoy

Grain is delivered to floating mills under the Grand Pont in Paris. [Bibliothèque Nationale, Ms. fr. 2092, f. 37v.]

the performances of musicians and acrobats, such as the tight-rope walker of 1389 who performed on a wire stretched between a tower of Notre Dame and the tallest house on the Pont-St.-Michel.

Inland waterways, virtually unknown in the ancient world, had their Western beginnings in the Middle Ages.[104] To improvement of river navigation by levee building, the eleventh century began adding construction of canals, for which, however, canal locks remained a Chinese secret. In their place, level-changing stations were created by erecting inclined planes with boat-carrying platforms powered by horse winch. Despite such efforts, the inland waterways carried only a minor part of the growing commercial traffic of the late twelfth century.

Cog, Compass, and Rudder

The commercial expansion manifested itself even more conspicuously in seaborne than in overland transportation. Ships, both sail and oar, multiplied in northern as in southern waters. Luxury goods were a prominent part of cargoes, though the term "luxury" meant different things in Europe's two regions. In the north, luxury goods included manufactures of any kind which had to be carried from their specialized places of origin to the rest of the area. In the Mediterranean, manufacture was more widely diffused and for the most part was traded locally; luxury goods here meant primarily the very expensive products of the Far East, laboriously conveyed to the frontiers of Europe by Arab ship and Asian pack train.

The most important eleventh-century development in northern waters was the improvement of the round-hulled cog. The flat bottom that had made the earlier cogs handy for beach landings had also made them prone to leeway—drifting sideways off course. As port facilities multiplied, it became practical to give the cog a round bottom and deep keel, substantially alleviating the problem. Entirely sail propelled, the new model carried a single huge square sail of cotton or linen canvas, roughly 190 square yards in area, which could be augmented in

Boatbuilder's tools: Noah building the ark. [Bodleian Library, Ms. Barlow, f. 53.]

fine weather by the addition of "bonnets," small pieces of canvas attached to the bottom edge. Flexibility was further increased by "reef points," short lengths of rope run through the sail in rows to roll up and tie the shortened canvas. Clinkering was kept but reversed, the bottom plank made to overlap the one above, giving a stronger hull. Freeboard (the side of the hull above water) was high, making the new cog distinctly an open-water overseas trader, economizing on manpower per cargo ton. Where an old-fashioned knorr of the type that colonized Iceland and Greenland could carry 50 tons with a crew of twelve to fourteen, the new cog, with its tublike proportions of a beam one third or more of its length, could carry 200 tons with a crew of eighteen to twenty. By the late twelfth century, cog capacity was reaching 300 tons.[105]

Surprisingly, the cog also proved an excellent warship and in doing so added a new element to ship structure. The height of its deck made it virtually impregnable to anything but another cog; to match its advantage, Scandinavian longships were given "castles," temporary platforms for archers. The cogs' builders promptly added castles to their own vessels, regaining the advantage, and the castle, in bow and stern, soon became a permanent feature.[106] Though the oar-propelled longship could maneuver circles around it, the cog's lofty castles gave its archers a commanding position. Bulky, ungraceful, competent, and formidable, the cog became the workhorse and war-horse of the German Hanseatic cities that by the twelfth century were beginning to dominate Baltic and North Sea commerce.

In the Mediterranean, too, ship styles evolved with changing conditions. For one thing, a new fashion in piracy emerged. As Muslim nests on the European coast were ferreted out and European ports given stronger defenses, the Arab corsairs turned from amphibious raiding to attacking ships at sea. The Europeans joined in the game by fitting out privateers of their own and turned attention to making their cargo carriers more defensible. The galleys that specialized in high-value cargoes were made larger, lower, and faster, and Venice pioneered the castle in southern waters. The round sailing ships were also enlarged. The tideless Mediterranean had always presented problems in beaching ships for loading and unloading, and as ship size grew, the need for port facilities pressed, stimulating growth of the major port cities—Genoa, Pisa, Venice, Amalfi, and others. Transport of pilgrims and Crusaders and logistical support of the Crusades reinforced the stimulus. "By 1250 most significant ports in both northern and southern seas had at least one [quay]" (Richard Unger).[107]

Both docks and ships were made of wood. Venice, with its piers and Arsenal, was the first medieval port to feel the pinch of forest depletion, especially the shortage of certain types of tree: oak for hulls; larch and fir for internal planking, masts, and spars; elm for capstans and mastheads; and walnut for rudders, one of two Chinese inventions that began appearing in the twelfth century.[108]

* * *

Rather surprisingly, the mariner's compass did not arrive in Europe courtesy of the Arabs, who did not use it for navigation until the thirteenth century. Joseph Needham conjectures that it traveled overland, via the Silk Road, and not in a navigational context but as an astronomical-astrological and surveying instrument.[109] In its earliest European form, the compass followed the Chinese prototype, a magnetized needle transfixed on a straw, reed, or chip of wood floated in a bowl of water. The Chinese also produced a dry suspension, with the needle pivoted on a bamboo pin set in a hole on a small board; a European version, perhaps independently developed, consisted of a circular box with the needle rotating on a vertical pin.

The first Western textual reference to the mariner's compass occurs, in the twelfth-century English scholar Alexander Neckam's book *De naturis rerum* (On the natures of things): "The sailors ... when in cloudy weather they can no longer profit by the light of the sun, or when the world is wrapped in the darkness of the shades of night, and they are ignorant as to ... their ship's course, they touch the magnet with a needle. This then whirls round in a circle until, when its motion ceases, its point looks direct to the north."[110] The date of Alexander's description, probably written in Paris, has been fixed at about 1190. At that early date the compass was by no means in universal use aboard ships.

The stern rudder, also long known in China, probably had an independent European provenance, either in Byzantium or, more probably, in the Baltic. At least in Europe, its adoption was apparently not motivated by the need for better steering apparatus, since the age-old steering oar still gave satisfactory service. However, the new, larger cog now rose so high in the water that it required an extremely long steering oar when heeled over, presenting serious problems for the steersman. The cog's straight sternpost offered a likely place to hang a rudder, manipulated at first by a tiller outside the hull, later through a square port cut in the stern. No marked improvement in steering was gained; a ship still depended on shifting sails to execute a radical change in direction.[111] But the new device saved time and effort by reducing drift and holding the vessel on course.[112]

By the late twelfth century, the two-masted ship had made its appearance in the Mediterranean. Both Venetian and Genoese shipyards turned out two- and three-deckers with a pair of masts carrying lateen sails, their large cargo holds reducing freight rates and so stimulating commerce in bulk goods.[113] Even with stern rudder and lateen rig, sailing in the Mediterranean presented problems. Prevailing winds and currents made outward voyages east and south from Italy much easier than the return. In 1183 Ibn Jubayr, an Arab of Spain, sailed on a Genoese ship from Ceuta, on the Strait of Gibraltar, to Alexandria in only thirty-one days, but the much shorter return voyage from Acre in Asia Minor to Messina, Sicily, took him fifty-one days.[114]

Galleys were handicapped in heavy weather because of their low freeboard, provided to afford maximum leverage to the oarsmen: the more nearly parallel the oars were with the water, the greater the mechanical advantage.[115] Under sail, the galley had an additional problem, being unable to heel over very far in tacking into the wind. Another twelfth-century traveler, Roger of Hoveden, wrote, "Galleys cannot, nor dare not, go by that route [open sea from Marseilles to Acre] since, if storms strike, they may be swamped with ease. And therefore they ought always to proceed close to the land." The danger, added to the need to rest the crew, generally dictated coastal routes for galleys, which could maneuver away from lee shores without difficulty. But open-water crossings gave the sailing ship a great advantage in carrying cargo, especially in the easterly and southerly directions.[116]

The Renaissance of the Twelfth Century

The innovations of the central Middle Ages in agriculture, power sources, handicraft production, building construction, and transportation were accompanied by dramatic developments in the realm of pure science. "The tenth century, though on the surface a time of invasion, cruelty, barbarism, and chaos," writes Richard C. Dales, "is nevertheless the turning point in European intellectual history in general and the history of science in particular."[117]

One of the Middle Ages' most important creations, the medical school, was founded at Salerno in the eleventh century, when by no coincidence the earliest cultural contacts with Islam occurred. General higher education had its beginnings in the cathedral schools founded in the tenth through twelfth centuries in Paris, Chartres, Rheims, Orléans, Canterbury, and other cities. Emphasis varied. Partly because of the Church's need to determine the dates of its movable feasts, astronomy was a favored subject, notably at Rheims under the scholar-teacher Gerbert, later Pope Sylvester II (reigned 999–1003).

Gerbert adopted the cosmology of Ptolemy as the most reasonable of those available and reintroduced two ancient but neglected devices for classroom demonstration: the abacus or calculating board, a set of counters arranged in columns for performing arithmetic, and the armillary sphere, a representation of the cosmos by an assembly of balls, rods, and bands. Gerbert may also have employed the astrolabe, known to have reached Christian Europe from Muslim Spain at about this time. The Arabs had improved the instrument, giving it what amounted to its final form, as a stereographically projected map of the heavens, with the stars and ecliptic marked on a skeletal plate (the rete) that rotated over another plate (the climate) giving local coordinates.

The cathedral schools' teaching was not for the clergy alone; by the twelfth century, some fathers enrolled sons to prepare them for careers in the law and other secular callings, including the growing governmental bureaucracies. The abacus, coming into wide practical use during the eleventh century, was introduced into the Norman-English exchequer in the twelfth.[118]

In the mid–twelfth century, the "precocious humanism"

(Carl Stephenson)[119] nurtured by Gerbert, his pupil Richer, and other scholars met and merged with another current, the growing importance of the professions of law and medicine, to create the first universities, at Paris and Bologna. From its beginnings, the University of Paris as well as its early offshoot, Oxford, articulated "productive ideas concerning nature as a fit subject of study." Scholars such as Peter Abelard (1079–1142) formulated "a new approach to the systematic study of science" (Tina Stiefel) even before the works of Aristotle became available in Latin.[120]

But it was the Muslim-assisted translation of Aristotle followed by those of Galen, Euclid, Ptolemy, and other Greek authorities and their integration into the university curriculum that created what historians have called "the scientific renaissance of the twelfth century." Certainly the completion of the double, sometimes triple translation (Greek into Arabic, Arabic into Latin, often with an intermediate Castilian Spanish vernacular) is one of the most fruitful scholarly enterprises ever undertaken. Two chief sources of the translations were Spain and Sicily, regions where Arab, European, and Jewish scholars freely mingled. In Spain the main center was Toledo, where Archbishop Raymond established a college specifically for making Arab knowledge available to Europe. Scholars flocked thither, headed by the prolific translator Gerard of Cremona, credited with seventy-eight works, several of them lengthy. To the Greek writings were eventually added many Arabic glosses and commentaries.

Some original Arabic works were also translated, such as the trigonometric tables of the mathematician al-Khwarizmi, given to the West by the enterprise of Adelard of Bath in 1126. By 1200 "virtually the entire scientific corpus of Aristotle" was available in Latin, along with works by other Greek and Arab authors on medicine, optics, catoptrics (mirror theory), geometry, astronomy, astrology, zoology, psychology, and mechanics.[121]

As the rival cosmic systems of Aristotle and Ptolemy came under scrutiny, their differences drew critical attention. From a medieval point of view, Aristotle's system made logical sense but was devoid of practical value, that is, in dating movable feasts. That of Ptolemy was based on hard-to-believe (and in fact incorrect) celestial mechanics but "saved the appearances," that is, accounted for the apparent movement of the planets.[122] Both authorities confirmed two basic assumptions, one right and one wrong: that the earth was a sphere, a fact observable in an eclipse of the moon and in the apparent descent of a departing ship, and second, that it was the center of the universe. This was a pervasive assumption, congenial to psychology as well as religion, since it centered the universe on mankind. It accounted satisfactorily for the paths of all the heavenly bodies except five recalcitrant planets, for whose apparently eccentric orbits Ptolemy had invented an elaborate explanation. For some reason, while Ptolemy's massive astronomical work, the *Almagest*, was translated, his *Geography* was overlooked by Europeans until the fifteenth century, despite its greater practical significance. Islam, on the other hand, knew it through Syriac translations and probably also in the original Greek text and based a number of important Arabic treatises on it.

A fuller picture of the earth's surface and its inhabitants gradually emerged in the tenth and eleventh centuries, when despite wars, piracy, and Crusades, peaceful travel increased. The Arab geographer al-Biruni (973–1048) asserted that "to obtain information concerning places of the earth has now become incomparably easier and safer."[123] Much of the Arab geographical lore was imported to Europe by way of Sicily, whose Norman conquerors of the 1070s, inheriting and tolerating a population of Muslims, Jews, and Christians, maintained contacts with the Islamic world and showed a marked interest in scientific learning. At the command of King Roger II, a distinguished North African scholar, al-Idrisi (1100–1165), assembled an academy of geographers to collect and evaluate information on boundaries, climates, mountains, rivers, seas, roads, crops, buildings, crafts, culture, religion, and language, combining the study of previous geographical works (including

Ptolemy's) with original research. Travelers were interviewed, the crews and passengers of ships docking in Sicily were interrogated, and expeditions, accompanied by draftsmen and cartographers, were dispatched to areas on which information was lacking. In a word, scientific method was applied to geographical research.

The resulting compendium was called *Nuzhat al-mushtag fi ikhtiraq al-afaq* (The Delight of One Who Wishes to Traverse the Regions of the World), or, more simply, *al-Kitab ar-Rujari* (Roger's Book). It contained a world map and seventy sectional itinerary maps representing the seven climates of the habitable world (according to Ptolemy), each divided longitudinally into ten sections, and a minute account of the regions thus illustrated. To accompany it, Idrisi constructed a great silver disk almost eighty inches in diameter and weighing over 300 pounds, incised with the limits of the climates and outlines of countries, oceans, rivers, gulfs, peninsulas, and islands. Written in Arabic and never translated into Latin, *Roger's Book* exerted little direct, but probably considerable indirect, influence on European thought through the mingling of Arabic scientific traditions with Norman and Italian maritime enterprise.[124]

The twelfth century also witnessed the tardy introduction to Europe of the second of the great "false sciences," alchemy, whose sister, astrology, had remained known and practiced in unbroken continuation since Roman times. Once regarded as a pair of fruitless medieval exercises in superstition and charlatanism, the two have gained stature with the maturing of the history of science. An early recognition of alchemy's value was voiced by Francis Bacon (1561–1626), who recounted the story of "the man who told his sons that he had left them gold buried somewhere in his vineyard; where they by digging found no gold, but by turning the mould about the roots of the vines, procured a plentiful vintage. So the search and endeavors to make gold have brought many useful inventions and instructive experiments to light."[125]

The first Arabic treatise on alchemy to be translated into Latin was rendered by Robert of Chester in 1144, quickly followed by several more as the new science caught on.[126] Alchemy had two aspects, theoretical and practical. The involved and mystical theorizing led nowhere, but the practice of alchemists in their laboratories became the direct ancestor of modern chemistry and chemical technology.

Practicing alchemists pursued two aims: the conversion of base metals into gold, usually by means of the elusive "philosophers' stone," and the discovery of the "elixir of life" (also known as the "most active principle" or the "fountain of youth"), which would confer immortality. The first kind of research, based on the hypothesis that gold is the sole pure metal and that all the others are impure versions of it, led to accumulation of knowledge about physical and chemical reactions, while the second kind gradually turned into iatrochemistry, the search for healing drugs.

Medieval alchemists, Arabic and European, introduced no wholly new equipment into their laboratories, but they created a multiplicity of furnaces and stills. Furnaces of varying sizes were needed partly to accommodate the diversity of fuels—charcoal, peat, dried dung—and partly to provide the varied temperatures required for calcination (reduction of solids to powder) of different substances. Bellows were much employed, causing alchemists in France to be nicknamed *souffleurs* (blowers).[127] A parallel collection of stills served the alchemists' other principal technique, distillation (boiling and condensation to separate compound substances).[128] The typical still was a tall vessel shaped like a church spire, mounted on a short tower; the fire in the lower part heated liquid whose steam condensed in the upper part and was guided by a long spout to another vessel.[129] Early stills lacked an efficient cooling device, and volatile liquids were usually lost. The still in which condensation was effected outside the still head may have been invented by a physician of Salerno named (for the city) Salernus (d. 1167). One product of the process, alcohol, strengthened by redistilling, found a variety of uses, as a solvent, a preservative, and the basis of brandy, gin, and whiskey, at first taken medicinally, later recreationally.

Both astrology and alchemy remained sources of interest to intellectuals long after the Middle Ages, but the importance of the magical element in medieval science has been exaggerated. "The striking thing about the [twelfth] century," in the words of Richard Dales, "is the attitudes of its scientists . . . daring, original, inventive, skeptical of traditional authorities . . . determined to discover purely rational explanations of natural phenomena," in short, portending "a new age in the history of scientific thought."[130]

The healthy skepticism of the men of the twelfth-century renaissance was underpinned by a distinct, even enthusiastic naïveté (Abelard and Héloïse, his bluestocking mistress, named their son Astrolabe). Devout clergymen, they innocently conceived investigation of the natural world as their Christian duty, undertaken in a spirit of gratitude toward God, "to help men reach a higher level of understanding of the Creator" (Tina Stiefel).[131] So far from anticipating conflict between study of natural phenomena and Church doctrine, they felt that their researches helped combat the ancient, still popular pagan superstitions centered on magical trees, rocks, streams, and forests.[132] In the demythologizing of nature, the medieval Church, following the lead of Boethius, anticipated the Renaissance humanists. As George Ovitt observes, "The *scientia* of the Middle Ages was theology, but theology was understood to include not only the nature of God and of moral laws, but also the nature of the world created by God."[133]

EUROPE 1200

Of all the changes in the appearance and activities of western Europe by the end of the twelfth century—castles and cathedrals, land clearance, swamp drainage, waterwheels and windmills, hospitals and universities—the most impressive lay in the realm of commerce. Many more pack trains and wagons were on the road; many more round ships sailed northern and southern seas. Behind commerce, industry flourished, an industry that still fashioned articles one by one, by hand, but an industry vigorous, growing, and with potential for the future.

In the old-fashioned history books with their political maps, Europe of 1200 figured as an incoherent jumble of petty principalities and cloudy sovereignties, seeming to stand still or even move backward in respect to the modern world. But looking past popes, Holy Roman Emperors, counts, landgraves, and archbishops, and focusing on the worlds of agriculture, commerce, manufacture, and intellectual activity, one can discern a region economically coherent and intellectually dynamic, borrowing, adapting, inventing, and synthesizing technology.

6

THE HIGH MIDDLE AGES

1200–1400

T HE TWO CENTURIES THAT FOLLOWED THE period of European emergence are usually pictured in sharp contrast: the thirteenth, sometimes called the Golden Century, an era of affluence and growth, the fourteenth one of catastrophe and contraction. The contrast, however, has been overemphasized at the expense of important elements of continuity.

Throughout the thirteenth century, Europe's technological advances continued in all sectors, sustained, among other factors, by an era of mild climate favorable to crops.[1] Communication between Europe and Asia benefited significantly from the conquests of the Mongols, whose ferocity in war contrasted with the peaceful character of the empire they imposed from Hungary to the Pacific Ocean. Papal ambassador John of Pian de Carpine (1246), friar William of Rubruck serving as an envoy from King Louis IX of France (1253), and merchant Marco Polo (1260) were only the most famous of the Europeans who were now able to make the direct acquaintance of Chinese civilization and technology. One aim of the European visitors, however, Christian proselytizing, had little success among either Mongols or Chinese.

In Europe, the Golden Century shone not only in Gothic architecture but in the rapid expansion of the Commercial Revolution. In its vanguard stood the cities of northern Italy, whose businessmen discovered (or borrowed) new machinery, new processes, and new business techniques. The fall of the Crusader states to Muslim reconquest had little effect on the predominantly Italian merchant colonies in the Levant ports, which continued to govern themselves and do business, now with the protection of the Islamic authorities. Europeans even moved into Egypt, where they had not previously ventured, to pick up the spices brought across the Indian Ocean by the Arabs, for which Alexandria was the chief entrepôt.[2] In the fourteenth century, the traffic shifted toward the territory of the friendly Mongol Empire, inspiring Europeans to think grandiose thoughts about gaining control of the whole spice trade. In 1318 a Dominican friar, William Adam, proposed stationing in the Red Sea a blockading squadron of galleys manned by Genoese, whom he esteemed as "the best and greediest of sailors," to shut the Arabs out of the spice trade altogether.[3] The scheme was not so much chimerical as premature, and in the meantime the Mongol Empire first turned hostile to European Christians and then collapsed, helping to deflect European attention toward the possibility of circumnavigating Africa.

The emerging Christian kingdoms of the Iberian peninsula, most strategically located for that enterprise, also turned to Genoa for shipbuilding and navigational help. Ugo Venta was the first of several Genoese admirals of Castile. Manuel Pessagno, appointed the first admiral of the Portuguese fleet in 1264, was succeeded by five generations of his family.[4] Genoese expertise was part of a substantial input of European naval and military technology to the Reconquista, which among other things signaled a shift in the pattern of technical diffusion between Christian and Islamic cultures: improvements in arms and armor were now copied by the Muslims from the Europeans.[5]

Among its far-reaching effects, the accelerating Commercial

Revolution provoked a demand for more coin metal, stimulating a historic development in central Europe: the opening of the rich silver-copper-gold mines in Bohemia, the Carpathians, and Transylvania, whither German miners from the Harz Mountains brought their skills. Underground mining developed rapidly, with the introduction of the vertical waterwheel for drainage, and the Chinese wheelbarrow.[6]

In northern Europe, the cities of the Hanseatic League not only successfully battled pirates but, accepting a challenge from the Danes, overthrew Danish hegemony and became the dominant power in the northern seas. Their larger ships reduced freight rates, and they built lighthouses and quays, marked reefs and channels with buoys, trained pilots to navigate coastal waters, and composed their own maritime law.[7]

The disasters of the fourteenth century came in three shapes: fi st, the bankruptcies of several of the great Italian commercial and banking houses; second, a series of wars, especially the English-French Hundred Years War; and third, a succession of famine years (1315–1317), followed a generation later by the terrible visitation of the Black Death, which crept across Europe in 1348–49, abating only to return at intervals in this and the following century.

The source of the Black Death remains a mystery (a contemporary chronicler attributed it to the Mongol siege of Genoese-held Caffa, on the Black Sea, where corpses of plague victims were catapulted into the Genoese compound), and even the disease's identity as bubonic plague has been questioned.[8] What is certain is that a European population that was already declining was devastated. Families were extinguished, villages left deserted, cities depopulated. Yet the resilience with which Europe survived and recovered is as noteworthy as the calamity itself. Agriculture and commerce resumed, property was redistributed among survivors, and earlier marriage lifted the birthrate, beginning the restoration of the population.[9]

In sum, the much-advertised disasters of the fourteenth century only temporarily disrupted economic life and had no discernible effect on technical progress, where the train of improvements continued in cloth making, construction, metallurgy, navigation, and other arts. On the political level, the century saw a shift toward modern political organization, especially marked in England and France, as central monarchies acquired stronger instruments of power, better sources of revenue, and expanded administrative machinery.

Perhaps most significant, if least obtrusive, was the advance toward modern business methods and organization. "Unstinting credit was the great lubricant of the Commercial Revolution," according to Robert Lopez.[10] The formation of large trading companies dealing extensively in credit transactions gave rise, first in Italy, then elsewhere, to commercial banking, dominated by such swiftly growing family-based institutions as the House of Medici. To serve the more complex business world, new record-keeping devices, notably double-entry bookkeeping, were invented.

The Countryside: Estate Management and the Black Death

The European countryside experienced the vicissitudes of the thirteenth and fourteenth centuries in the most marked degree. The thirteenth was a period of intensive cultivation, from which most of our information about medieval estate management derives; the fourteenth one of crop failures, plague, and agrarian disorders, and subsequent adjustments.

Thirteenth-century agriculture produced little new technology but a change in management methods and a substantial increase in production. Many lords had previously been content to "farm" their estates, that is, turn them over to outside entrepreneurs who paid a fixed yearly sum and collected rents, fines, and other proceeds. As the market grew with the population, opening the way to cash profits, the lords tended to assume direct control. Their tenants also benefited from prosperity and the money economy to buy their way free from many of the old manorial obligations. A few acquired land and even got modestly rich.

A sign of the lords' new interest in their agricultural affairs was the appearance of medieval Europe's first practical treatises on agronomy. Columella, Varro, and other Roman writers had long been read in the monasteries, and a few Arabic works from Spain had been translated, but the value of both Roman and Muslim authors was limited by their focus on Mediterranean-style farming methods, designed for conditions quite different from those of northwest Europe. In the second half of the thirteenth century, a number of treatises appeared, written in the vernacular and addressed to contemporary estate owners.[11]

One of the most influential, the *Husbandry* of Walter of Henley, was written by a former English bailiff (manorial officer). Walter began with some Poor Richard advice on the need for prudence, forethought, and honesty ("He who borrows from another robs himself," "Put [your] surplus in reserve," "Have nothing from anyone wrongfully"), then covered every aspect of agricultural management: surveying and evaluating demesne, pasture, buildings, gardens, woods, tenants, yields; selecting stewards and bailiffs and overseeing laborers; plowing, sowing, drainage, seeding, marling and manuring, dairying; raising of sheep, pigs, and poultry; and, a significant new element in agricultural management, the keeping of accounts.[12]

"The manorial account in the form historians know it is a late twelfth- or thirteenth-century innovation," according to Michael Postan.[13] Drawn up at the end of the agricultural year by the reeve, a villein often elected by the villagers themselves, the accounts painstakingly detailed receipts, expenses, stores on hand, and stock. The illiterate reeve kept track of the figures by marks on tally sticks, which he read off to the lord's steward or clerk. Formal accounts came to be adopted on most estates, always following the same pattern, the uniformity suggesting the quick spread of information among administrators.

The emphasis of the new agricultural treatises was on conservation—maintenance of yields, protection of livestock, avoidance of waste—rather than on increasing output. The bias suited the lords to whom the advice was addressed. Like Roman landlords, they had little interest in effecting capital improvements. They were essentially consumers rather than producers, and consumers on a liberal scale, whose openhanded generosity toward retainers, staff, and guests enhanced their reputations, giving the impression that "the springs of wealth were inexhaustible" (Georges Duby).[14] Yet Walter of Henley and the other authors of treatises encouraged improvement of livestock breeds ("Do not have boars and sows unless of good breed") and of seed ("Seed grown on other ground will bring more profit than that which is grown on your own"),[15] and at least to some extent they were heeded. The great ecclesiastical estates in England imported breeding stock from the Continent, and thirteenth-century lay lords made capital investments in animals, land, tenants' houses, barns, bridges, mills, and fishponds. Henry de Bray, a petty landholder in Northamptonshire, widened a stream to provide a fishpond and built a mill and a bridge. Benedictine abbot John of Brokehampton built sixteen water mills and a number of windmills on his large estate.[16]

With affluence and progress came the first recognition of limits to growth. Increasing population dictated increase of the cultivated area. The resulting impingement on the wilderness, combined with the growing pressures of construction and industry, brought Europeans for the first time to a consciousness of the forest's limits. Royal and seigneurial regulations curtailed land clearance and tree cutting, as well as restricting other activities less obviously harmful to the forest—grazing animals, gathering nuts, and collecting deadwood for fuel. The more valuable trees, especially beeches and oaks, were objects of special protection. Despite such measures, in the fourteenth century French forests had diminished by more than half since the time of Charlemagne, while those of England had fallen by a third from the Domesday accounting.[17]

Growth was already slowing down and the era of prosperity coming to a halt when the first disaster of the fourteenth century struck: famine, following two successive harvest failures brought on by bad weather (1315–1317, possibly part of a long-term climatic change), and epidemics of murrain and cattle disease that infected flocks and herds. The poorest households were the hardest hit. Faced with bad times, many lords turned back to farming out their demesnes, once more becoming absentee rentiers, and in England the beginnings of a major shift appeared, away from crop farming and toward sheep grazing.

In 1347–48 a countryside already weakened by famine was visited by the Black Death, which left thousands of holdings vacant, temporarily crippling the manorial system by making it impossible to collect rents or enforce labor services. Surviving manorial records are tersely grim: "Rent lacking from eleven cottages . . . by reason of the mortality." "Three capons and no more this year because those liable to chevage are dead." "Of divers rents of tenements which are in the hand of the lord owing to the death of the tenants."[18] Yet within a year, life had returned to the appearance of normality. In the Midlands manor of Halesowen, where the plague had struck in May 1349, killing at least 88 out of 203 male tenants and wiping out some households, a modern study reports: "The records of the [manorial] court held between August 1349 and October 1350 show that the villagers harvested their crops and pastured their animals. They married and bore children in and out of wedlock," and in short carried on business as usual. Vacant tenancies were taken up by sons and daughters, wives and brothers, or other relations, the average size of holdings increasing as land that might previously have been divided was passed intact to a single heir.[19] The price of land fell, and the pressure on the forest relented, to such effect in England that total forest area thenceforth remained stable until the nineteenth century.[20]

In some places, notably England, the Black Death contributed to social unrest as manorial officials, backed by a royal "Statute of Laborers," sought to enforce work services more strictly on surviving tenants. The addition of heavy war taxes stimulated the Peasants' Rebellion, of 1381, only one of a number of European outbreaks. "A chain of peasant uprisings clearly directed against taxation . . . exploded all over Europe" (Georges Duby).[21]

There was a deeper-seated cause of the rebellions than taxation, noted by Shakespeare in a much-quoted line put into the mouth of one of the rebels of 1381: "The first thing we do, let's kill all the lawyers" (*Henry VI, Part II*, IV, i). Lawyers were indeed killed and manorial records destroyed in revolutionary violence aimed against the institution of serfdom. The revolts were suppressed, but as often happens with revolutionary movements, their aim was attained in the aftermath, as over the course of the fifteenth century serfs and villeins succeeded in buying freedom from, or simply refusing to pay, the old servile dues.

In the last half of the fourteenth century, peasant holdings grew while the abandonment of some arable land provided more pasture and stimulated an increase in livestock, which in turn provided more manure and probably benefited crop yields. Wealthy townsmen entered into sharecropping arrangements with peasants. "The conduct of village economy," says Georges Duby, "passed decisively into the hands of peasants backed by townsmen's money."[22]

Labor-intensive one-crop grain cultivation, to which both lord and peasant had clung almost superstitiously, retreated in many areas, its place taken either by land-intensive stock raising, which supplied wool, hides, meat, and cheese for the market, or by diversified fruit and vegetable farming, also for the market.[23] A more resilient agricultural economy gradually emerged, less vulnerable to the disasters visited on cereal-crop farming.

Cloth, Paper, and Banking

By the thirteenth century, the Flemish wool cloth industry had moved out of the villages and into the towns of the Scheldt valley, where a new organization of production appeared, to spread presently to Italy, England, and southern Germany, and to survive into early modern times, particularly in eastern Europe. This was the putting-out system. The cloth merchant, already strategically situated as middleman between weavers and the market, now took over the role of middleman between the weavers and their source of supply, the English sheep farmers. From there it was only a step to make himself the entrepreneur in what has been called "a factory scattered through town."[24]

A unique document from thirteenth-century Douai gives an intimate picture of the putting-out system at work. The record of a legal proceeding in 1285–86 against the estate of Sire Jehan Boinebroke, cloth merchant and notorious skinflint, by forty-five clothworkers and other claimants illuminates the human as well as the economic aspect of the system. Boinebroke contracted through his agents to buy wool from Cistercian monasteries in England, making a down payment of about 3 percent. When the wool arrived, he sold it to the weaver, who took it home to sort, card, spin, and weave, with the help of his wife and children. The weaver then sold the unfinished cloth back to Boinebroke, who sold it to a fuller for cleaning and treating, after which he bought the finished cloth back and either sold it to a dyer or sent it to his own dye shop behind his house. Finally, he sold the fulled and dyed cloth to his agents, who took it to sell at either the Douai cloth market or the Flemish or Champagne Fairs. Thus Boinebroke bought and sold the wool four times. A sack of fleece that cost seven pounds in England might sell as cloth for forty pounds in Champagne, with a large proportion of the markup going to Boinebroke, who was protected at all stages of the transaction against market fluctuations and other reverses. If war interrupted traffic to the Champagne Fairs, he could simply refuse to buy back the cloth at any stage in its manufacture, or could buy it back at a low price.

Boinebroke was a landlord as well as an employer, renting whole streets of houses to his workers in the old lower town of Douai and the marshy area between the lower and the upper town. He was also a recurrent member of the patrician town council, which elected its own successors every fourteen months and governed without apology in its own class interest.[25]

The claims against Boinebroke in the lawsuit show him to have been a grasping and heartless tyrant in his dealings with the clothworkers, though the fact that the suit could be brought—and a third of the claims honored by the court—

shows that justice was not entirely lacking in the cloth cities. The main revelation of the document, however, supported by information from other sources, is that the age placed few restrictions on the exploitation of labor. The result was endemic class warfare in Flanders and Italy.

In all the disturbances, the key figure was the weaver. The most important component in the industry, he employed a loom that was the only complex implement involved in the many steps of the cloth-making process. The chief adversary of the merchant-entrepreneur, he was the first worker in history to bring an industry to a halt by going on strike (in Douai, in 1245).[26]

An important further step in the mechanization of the cloth industry was signaled in the late thirteenth century by the introduction into Europe of the spinning wheel, which may have originated in the Near East or India (the earliest clear illustration is from Baghdad in 1237). Arnold Pacey believes that "the westward dissemination of the silk and cotton industries may have stimulated local responses . . . Some form of wheel for spinning may have been suggested to the minds of a number of individuals in quite different places."[27] In its earliest Western form, it consisted of a small spindle mounted on bearings and connected by a belt to a large wheel. The spinner held the mass of fiber on a distaff in her left hand, imparting a twist as with her right she fed it to the spindle, which she kept in

Two thirteenth-century innovations: the spinning wheel, left, and the carder with metal teeth. [British Library, Luttrell Psalter, Ms. Add. 42130, f. 103.]

rotation by intermittently giving the big wheel a turn. The invention is noteworthy for its early embodiment of belt transmission and its use of the flywheel to maintain a steady flow of power.

Resistance to the new device in the wool cloth industry surfaced immediately, principally from the merchants, who saw it as impairing quality by producing thread that was rough and uneven. Lacking the pedal and flyer that were added later, the early spinning wheel was not well adapted to produce an even thread. The earliest documentary evidence of its introduction in the West comes from statutes of the drapers' guilds in the 1280s banning its use. Nevertheless, the machine gradually came to be employed under the spur of the chronic imbalance between spinning and weaving. Several hand spinners were needed to supply one weaver; the wheel roughly halved the number.[28]

The spinning wheel was at first reserved for the spinning of weft, which did not need to be as strong as warp. "The spinner . . . much values her thread which was spun on the distaff," says a verse in the *Livre des mestiers* (Book of crafts) of Bruges (c. 1340), "but the thread which is spun on the wheel has too many lumps and she . . . earns more to spin warp at the distaff than to spin weft with the wheel."[29] A fourteenth-century document of the Florentine Arte della Lana, giving detailed instructions for the preparation of wool for weaving, directs that the long fibers left after combing should be sent to the country to be spun on a spindle for the warp, the short fragments spun on the spinning wheel for the weft.[30]

Warp and weft were usually spun in opposite directions, clockwise (Z-spun) for warp and counterclockwise (S-spun) for weft, so that a merchant or weaver could tell at a glance whether yarn was intended for warp or weft. Meanwhile, spinning remained the most poorly paid occupation in the cloth industry, and spinners (all female) were not included in the craft guilds.

Another innovation in wool cloth manufacture made its appearance in thirteenth-century France and Flanders: a carding instrument with metal teeth, to replace the old bone or wooden wool comb in preparing fibers for spinning. Again, the innovation met with resistance because of the threat to quality—the device saved time, but the finished product was slightly rougher, with shorter fibers. Little by little, as techniques were improved, metal carders came into wider use, first with the weft, later with the warp. The fibers were oiled or buttered, one part of them attached to the card held in the right hand, the other card drawn across it, an operation that was repeated eight to ten times. Carding was usually done by the spinners.[31]

Still another textile invention was the toothed warper, for preparing bundles of warp threads of equal length to be placed on the loom. A square frame that leaned against the wall, the warper was armed with rows of pegs, around which the warp, drawn from a dozen or more large bobbins turning on an iron rod, was wound in a zigzag pattern. The instrument made it possible to use long warp threads, producing long pieces of cloth, the lengths standardized in each city so that a buyer knew the dimensions of a bolt of cloth simply by its city of origin.[32]

The horizontal loom of the twelfth century could produce

A toothed warper for preparing bundles of warp threads to be placed on the loom. Threads are drawn from bobbins at lower right and wound around rows of pegs on frame that leans against the wall. [Drawing adapted from the fourteenth-century Kuerboek of Ypres.]

cloth only as wide as the weaver could reach on either side to pass the shuttle through; in the thirteenth century a wide horizontal loom appeared, operated by two weavers, who passed the shuttle back and forth between them. The vertical warp-weighted loom at last withdrew to Scandinavia, the Faroe Islands, and Iceland, where it remained in use into modern times. The vertical two-beam loom became a specialized device for the weaving of tapestry, an ancient art that gained sudden popularity in the fourteenth century and flourished all over Europe in the luxury workshops that produced artifacts for princes. In tapestry weaving, the weft was not carried all the way across but worked by hand back and forth in each color over limited areas in a process similar to darning.[33]

More surprising to historians than its slow adoption of the spinning wheel and the metal card is the textile industry's failure to apply waterpower. In eighteenth-century England and America, power was applied to cloth manufacture by waterwheels in no way different from those available in medieval Europe. Robert Lopez, noting that silk throwing in Lucca was waterpowered, gives this explanation: "Wool yarn . . . was coarser and cheaper; there was no incentive to invest in a costly machine while it was possible to put out the wool to underpaid spinstresses." Fullers were better paid and consequently were provided with waterpower to increase their productivity.[34] However, it should be noted that the long, continuous filaments of silk present a different problem from the short fibers of wool, cotton, and linen, and that the silk process that was waterpowered was not spinning—consolidating short fibers into a single strand—but throwing, that is, twisting two or more of the long natural filaments together to make a stronger, heavier thread. The waterpowered silk-twisting mill could not be used with other types of yarn, which had to await a number of eighteenth-century inventions.

As the market for wool cloth grew steadily through the thirteenth century, England ceased to be merely a giant sheep ranch for Flanders and began its long history as a cloth-making center, while Florence expanded its industry from dyeing and finishing to total manufacture. At the same time that luxury wool cloth was being made for long-distance commerce, however, garments worn by ordinary people were still spun and woven locally, in town and country, by old-fashioned methods.

Other branches of the textile industry besides wool manufacture flourished in the high Middle Ages. The technique of knitting may have been brought from the Near East by pilgrims or Crusaders. Similarly, the notorious Fourth Crusade that stormed Constantinople in 1204 may have captured for Venice the secrets of silk culture and processing. Lucca and Palermo had already acquired the technology; now the Italian silk industry expanded rapidly. Sometime in the thirteenth century it acquired the Chinese silk loom and shortly after made the leap to waterpower, at almost the same time as China.[35] In the fourteenth century, Lucca had a silk mill employing 480 spindles driven by an undershot waterwheel.[36] Silk weaving spread to northern Europe, using raw silk imported from Italy. In Paris and London, women workers dominated the industry.

In linen manufacture, a simple device invented in the Netherlands in the fourteenth century speeded the preparatory process: the flax breaker, or hackling board, consisting of two parallel boards on edge, hinged to a third board that slammed down on a bundle of stalks laid across them.[37] At the same time, the spinning wheel had a more immediate impact in the linen industry than in the woolen. The combination of the two inventions brought on a large increase in the production of linen shirts, bed sheets, undergarments, towels, and coifs.[38]

Meanwhile, in the thirteenth century the cotton industry of northern Italy had grown to a position rivaling that of wool in number of workers and size of capital investment. Cotton was the only major export industry manufacturing low-priced goods for popular consumption, with profits dependent upon a large volume of turnover. As a result, it developed a unique system of organization, with regional subdivision of labor and standardization of products and implements.[39]

The earliest centers of cotton production in northern Italy

were Milan, Piacenza, Pavia, and Cremona. In the thirteenth century the industry spread to other cities via organized migrations of skilled workers, who brought their techniques in return for tools, rent-free shops, interest-free loans, and rights of citizenship. The cities monopolized the most advanced processes, such as beating, weaving, stretching, dyeing, and finishing, which required full-time labor of professional craftsmen, leaving to part-time rural workers the less demanding procedures of spinning, warping, bleaching, and fulling.[40]

The spinning wheel met none of the resistance in cotton manufacture that it faced in the wool industry, since its use did not affect the evenness and fineness of the thread but rather contributed to its uniformity. Increasing the speed of yarn-making threefold, the spinning wheel made it possible to turn out large quantities of cotton thread of standardized weight, and, as an end product, cloth based on warp and weft threads of prescribed weight and quality.

Cotton looms were of standard dimensions, as were loom reeds or beaters—combs used to form the weft threads into a straight line and keep the warp threads evenly spaced. The vertical warper used in the wool industry also aided in the standardization of cotton, permitting the production of warp threads of fixed length, number, and weight, which were sold in skeins or sacks in the cities of northern Italy and could be mounted on any of the standard looms.[41]

The bulk of cotton cloth production was of light- to medium-weight cloth for undergarments, bedding, and summer clothing, competing with coarse linen. Linen was more durable but harder to care for and lacked some of cotton's visual and tactile qualities. For clothing and blankets, flannelettes and quilted cottons competed with cheap woolens. The Italian industry never produced the luxury cotton cloths of Islam—printed designs, tapestry weaves, brocades—but concentrated on production for the mass market.[42]

Some cotton cloth was simply bleached, but much was dyed, using special techniques developed in northern Italy. In the late Middle Ages a fashion change brought a demand for darker tones. Where white had long been the traditional color for mourning in the Mediterranean countries, in the fourteenth century black supplanted it in Spain and Portugal, at first for the court and the nobility but soon for all classes. The fashion spread to other countries, and later dark colors became popular for all kinds of clothing.[43]

Until the late thirteenth century, blue dyes were produced entirely from woad, native to Europe. Indigo, made from an Indian plant, produced a more brilliant and concentrated color, but the insoluble form in which the Indians exported it baffled European dyers, although painters successfully used it as a pigment. The problem was solved by the dyers of Venice with the help of Marco Polo, whose book contained a description of the method of preparing indigo that he had observed firsthand in India.[44]

A new style stimulated by the production of cotton cloth was the short, quilted, tight-fitting jacket or doublet introduced in Italy in the twelfth century as a garment for both sexes and spreading throughout the Continent and to England in the thirteenth. At first worn under a loose-fitting tunic, in the fourteenth century it became an outer garment, censured by the clergy for its briefness and tight fit. Cotton was also used for accessories, such as coifs, veils, wimples, handkerchiefs, purses, and linings, and, to meet another fashion dictate, stomachers, pads used to emphasize the female abdomen.[45]

The shipping of raw cotton, lightweight and bulky, presented a special problem. In the early fourteenth century, a method was invented for packing it more tightly—"cotton screwing," using a press or screw jack to cram as many sacks as possible into the ship's hold. The danger of weakening the ship's timbers led to regulations by the Venetian government limiting the proportion of cotton that could be packed into a vessel. Cotton shippers were able to offer lower freight rates to heavy accompanying cargoes, such as wood ash, salt, lead, and alum, and a system of differential rates was developed to balance loads, with lightweight commodities such as cotton paying double the rate of spices and four times that of heavy goods needed for ballast.[46]

*   *   *

One fashion change had repercussions in an entirely different industry. The popularization of the linen shirt and undergarments provided a bonanza of raw material for paper manufacture. The West acquired the complete papermaking process from China but, lacking the bark of the paper mulberry tree, used rags, especially linen. As linen's production and uses widened, more discarded rags became available, and as paper production rose, price declined and market expanded.

Manufacture was a two-stage process. In the first, rags were torn up by a rag cutter, soaked in a "rotting room," shredded, and beaten in troughs with spiked mallets.[47] In the second, the pulp was transferred to a vat of warm water, stirred, and immersed in a rectangular mold with a wire latticework bottom. Lifted out with a shaking motion, the layers of pulp were arranged in a pile, with felt separating the sheets, and the water was squeezed out. The sheets were hung to dry, then rubbed smooth with a stone, and finally plunged into a vat of sizing composed of gelatin and alum.[48]

Paper had been sized in China as early as the third century A.D., with the addition of gum or glue and later starch to prevent the ink from running. Writing with pen instead of brush demanded a stiffer sizing, introduced by paper manufacturers in Baghdad as the product migrated westward. Waterpower, also first applied in Baghdad, migrated west to Damascus by 1000 and by 1151 was used in mills in Moorish Spain.[49] The first paper mills in Christian Europe to apply it were in Fabriano, Italy, in 1276, where the watermark was pioneered six years later.[50] From Italy the process spread to France and Germany and, by the fifteenth century, to the Netherlands and England.

While in China paper found a variety of applications, in Europe its primary role from the first was as writing material. Book production had moved out of the monasteries in the twelfth century as commercial stationers began serving the university faculties and the mendicant orders, employing copyists (often former clergy) on a putting-out basis.[51] As the price of paper fell, the scribe became the largest cost factor in the production of books. Thus the advent of a mass-production writing material, in the context of an information-hungry world, supplied a powerful stimulus to the invention of a mass-production copying technique.

The expansion of the textile industry had yet another far-reaching effect as new forms of mercantile enterprise evolved among the Italian firms engaged in it. Temporary partnerships and joint stock companies had long been used in Italy to spread the risks of overseas trading. Instead of entrusting all his venture capital to a single ship, a merchant could put it into a joint company that divided the risks among several ships. The new *compagnia* (company) that came into prominence in the thirteenth century, however, was more than a temporary arrangement. At first a family partnership of father and sons, or brothers, people who lived in the same house and broke bread together (whence the term *cum pane*, with bread), all the partners contributed capital and all participated in management. Each accepted responsibility to third parties for debts contracted by the others, and all shared in the profits. In time the company came to include outsiders, but control remained with the family.[52]

The arrangement made it possible for merchants to stay at home while maintaining permanent branches, manned by "factors," in the great commercial centers: Bruges, Paris, London, Avignon (location of the papal court through most of the fourteenth century), in other Italian cities, and in Constantinople. A courier service between company headquarters and branch offices developed in the second half of the fourteenth century into the *scarsella*, a postal combine with regular weekly departures from Florence to Avignon by way of Genoa. The scarsella delivered the letters of its own members first, then several hours later letters of the general public, in other words, those of business competitors.[53]

Primarily dealers in wool cloth, the great companies also sold silks from Persia and China, pearls from the Persian Gulf and

Ceylon, tin from Cornwall, Polish and Scandinavian copper (imported via Bruges), lead from Sardinia, and armor from Milan and Germany.

The complexity of their operations demanded new methods of record keeping. Special records were kept for foreign customers, for dyeing and finishing establishments, for associates, petty cash, and inventories of stocks and equipment. Daily receipts and expenses were entered in a rough copybook, eventually to be transferred to a more systematic "great book." The company also kept a "secret book," containing the private accounts of partners and staff, their deeds of partnership, and the details of each partner's share.[54]

In the earliest surviving Italian account book, dating from 1211, memoranda were arranged in chronological order, with no separation of debit and credit. A little later, debits were entered in one part of the book and credits in another. Still later, the debits and credits of each account were presented on facing pages, a system known as "Venetian style." The evolution culminated in either Florence or Genoa sometime before 1340 in double-entry bookkeeping, in which each transaction was analyzed in terms of its effect on the assets on the one hand and the liabilities and owner's equity on the other. Every purchase and sale was entered twice. A purchase of cloth might be entered on the left-hand (debit) side of the ledger as an acquired asset and on the right-hand (credit) side as an expenditure of cash, a liability. The two sides of the ledger were always in balance (equal). At any point, subtracting liabilities from assets determined the owner's equity, allowing a company to keep day-to-day track of its fortunes.[55]

The evolution of bookkeeping can be traced in the extensive surviving records of the great Prato merchant Francesco Datini, from his early business dealings in Avignon in the 1350s to his death in 1410. Inscribed on the first page with a formula such as "In the name of God and of profit," or "In the name of the Holy Trinity and of all the Saints and Angels of Paradise," the Datini books at first were divided into sections for debits and for credits, containing in addition the novelty of the "impersonal account," representing such elements as office or administrative expenses. Losses and profits were recorded in some detail: "Here will be entered, God forbid, losses incurred on merchandise: 2 loads of wax, which Francesco di Boncorso bought for us at Genoa as shown on page 342. 2 florins, 7s. 6d." "Profits on merchandise will be entered here, God grant us health and profits. Amen. For profits on leather and sugar sold . . . the account is on page 174. 12 florins, 12 s." Later the accounts were drawn up Venetian style, and finally, from 1386 on, Datini's company began to use double-entry bookkeeping. By 1400 all the Datini companies were using the system, which in the course of the following century spread through Italy and Flanders, though elsewhere in Europe it remained unknown.[56]

The international character of the great companies' affairs led them inevitably into banking. The "bill of exchange," developed in the fourteenth century, was at once a way of supplying money to someone in another country and another currency and, like the old cambium contract, a form of loan in which the interest was concealed in the rate of exchange, thus evading the Church's condemnation of usury. In addition to issuing and accepting bills of exchange, for which they exacted a commission, the Datini company offered banking services that included letters of credit, loans to merchants, and many services to businessmen. Primitive examples of checks have been found in the Datini archives, although this instrument did not come into general use until the sixteenth century, money being usually withdrawn or transferred by verbal order in the presence of the parties involved. Bankers and money changers often had accounts with each other in a kind of forerunner of the modern clearinghouse, making it possible to transfer credit from one person to another even when accounts were in different banks.[57]

In the late thirteenth and early fourteenth centuries, some of the Italian banking companies became involved in public finance in England, with disastrous results. Backing Edward I in his conquest of Wales, the Riccardi company of Lucca was driven into bankruptcy. Several Florentine companies, financing Edward II's expenses, Edward III's wars with Scotland, and the first battles of the Hundred Years War, found themselves facing a similar fate. When their problems were aggravated by heavy loans exacted by the Florentine government to pay for its own wars, the companies were provoked to conspire in a coup d'état, whose failure was followed by a series of resounding bankruptcies. The lesson was taken to heart. In the following century, when bankers provided credit to governments, they made sure to attach adequate safeguards and were rewarded with suitable profits.

The Medieval City

In the prosperous thirteenth century, European cities began for the first time to rival in size and importance those of the classical world and contemporary Asia. Paris, London, Ghent, Bruges, Cologne, Florence, Genoa, Pisa, and others now sheltered behind their battlemented walls large and growing populations of craftsmen and merchants living lives free from feudal subjection, if not from modern tax oppression. It has been calculated that in 1380 half the population of Flanders and neighboring Brabant dwelt in cities.

In contrast with that of a late Roman or early medieval administrative center, the life of a thirteenth-century commercial and industrial city was full of activity: craftsmen working, merchants trafficking, wagons creaking, all the noise, bustle, and vitality of urban life. Its consumption was satisfied by a busy transport system, its exports and imports were underwritten by sophisticated credit arrangements, its many needs so successfully satisfied that one modern historian asserts that "few significant refinements were added" until recent times.[58]

That is not to say that no further "refinements" were needed. The high population density was met in part by houses sharing party walls and subdivided into small apartments.[59] Inevitably, problems of waste disposal and pollution arose. Tanners and butchers discarded entrails, blood, and hair in the streets; animals dropped manure; pigs, dogs, and rats raided garbage; open ditches served as sewers for storm water and wastewater; privies and cesspits occupied backyards. Traffic—horse, cart, pedestrian, and animal—crowded the streets, piling up at the gates where tolls were collected. Collisions provoked a stream of law-suits.[60] Heating and cooking, as well as industry, added smoke to the atmosphere. The smoke was almost entirely from wood and charcoal, whose fires had two other drawbacks: in combination with timber framing and thatched roofs, they created a citywide fire hazard, and they depleted the neighboring forest. Charcoal was especially wasteful of medieval man's best resource; while it gave more heat, essential in most industrial processes, its preparation burned up several times its weight in wood. Yet even where superior heating capacity was not needed, charcoal was often used because its lightness made it more transportable. Home heating was in any case extremely inefficient owing to the lack of window glass or insulation.[61]

Coal was known in Europe at least by the thirteenth century but was sparingly used out of fear of the toxic nature of its fumes. In England it was first gathered from outcroppings washed ashore on the northeast coast and was known as "sea coal," a name that stuck even later when it was mined inland.[62]

By the late Middle Ages, strenuous efforts to alleviate some of the problems were being made by city authorities, rich and influential men who unlike their modern descendants lived in the city themselves and had a direct interest in the environ-

Paving, from the fifteenth-century Chroniques de Hainaut. *[Bibliothèque Royale, Brussels, Ms. 9242, f. 48v.]*

mental quality. Two keys to urban sanitation were street paving and storm sewers, both of which were known to Rome and a few other ancient cities. Moorish Cordova paved its principal streets in the ninth century, but Paris and the largest Italian cities followed only in the late twelfth and thirteenth. Paving was indispensable for street cleaning, but besides being expensive to install, it needed endless upkeep. Cobblestone or brick surfaces had to be repaired and replaced under the pounding of heavy cart wheels that were either iron shod or, worse, wooden but studded with nails. Street repair was often done directly over the old broken surface, causing a rise in street level.[63]

Paris dug the first storm sewer in the fifteenth century and was copied by a few other cities, but at the end of the Middle Ages most towns still depended on open ditches that flooded in heavy storms. Systems designed to handle domestic sewage and industrial waste awaited the nineteenth century, when London pioneered a combined system. Meanwhile cities were still pocked with private cesspits, periodically emptied at "an understandably high cost" (Christopher Dyer).[64] Archaeologists found one medieval London latrine to contain a thousand gallons of ordure. Bylaws and building regulations sought to control maintenance and cleaning of the pits.

City water supply nearly always depended on local sources: wells, springs, and rivers. Professional water carriers assisted distribution from the fountain in the town square or the public well, served by bucket and windlass or bucket and counterweight. Better-off households had their own wells or cisterns, for which the proximity of cesspools and latrines posed chronic pollution problems and contributed to epidemics.[65]

Running water and domestic plumbing were not unknown in the Middle Ages but were limited mainly to monastic precincts, such as the cathedral priory at Canterbury, where water was carried by underground pipes to the infirmary, the refectory, the kitchen, the bathhouse, and the prior's chambers. After use, the wastewater ran off by a drain that flushed the "necessarium" (latrine). At Clairvaux, similarly, as described by St. Bernard's biographer, water was first channeled into a series

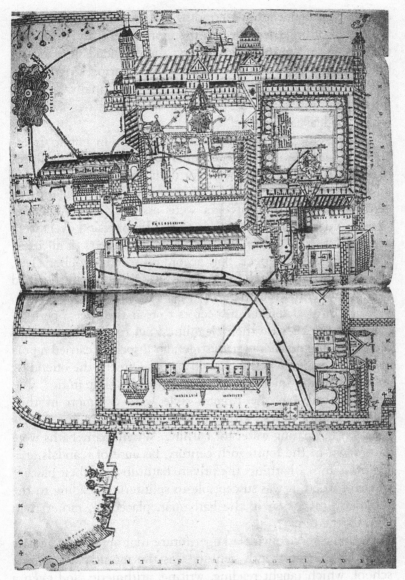

Water system at Canterbury. Water was carried by underground pipes from the piscina, or pool (upper left), to the infirmary just below it, the bathhouse (below the infirmary), the kitchen and refectory (right center), and the mill (bottom), before serving the necessarium, *or latrine (center).* [Trinity College, Cambridge, Canterbury Psalter, Ms. R 17, 1, f. 258.]

of industrial applications—grinding the grain and shaking the flour sifter, filling the boiler for the monks' beer, and operating the fulling and tanning machinery; then divided into several branches for cooking, washing, watering, rotating, or grinding, "always offering its help and never refusing," finally, "to earn full thanks and to leave nothing undone, it carries away the refuse and leaves all clean."[66]

In some cities, garbage disposal was handled by public street-cleaning services, usually on Saturdays.[67] Elsewhere, ordinances made householders responsible for their own rubbish, probably an ineffective solution. More successful was the regulating of certain occupations. Butchers were assigned waste-dumping sites or ordered to dump outside town. Tanners and dyers were usually restricted to the city's outer limits. Results of all these measures were imperfect, but, according to Christopher Dyer, towns of the later Middle Ages were "less filthy" than they had been a few centuries earlier.[68]

Some cities delegated inspectors to tour the streets periodically, not to check on their cleanliness but to detect encroachments. Riding down a narrow street, an inspector carried a pole across his saddle; where he could not pass freely, the offending shop owner was fined and forced to retract his shop front.[69]

Public baths in the Roman style were common in thirteenth-century cities, with the wall fireplace finding a new function in heating water for bathing. When many baths were shut down in the fourteenth century, because of scandals arising from unisex bathing, the private bathtub took their place.[70] Made of wood, it was susceptible to splintering, leading to the subsidiary invention of the bath mat, placed in, rather than next to, the tub.

A public service with a larger future that appeared in many cities by the fourteenth century was the municipal grammar school, which taught reading, writing, arithmetic, and even a little Latin to the sons and occasionally the daughters of merchants and artisans. The increasing literacy of the public widened the demand for books, now copied by professional scribes and marketed by professional booksellers.[71]

Crafts clustered in streets that were commonly named for them, though without benefit of street signs: Goldsmiths Street, Tanners Street, Shoemakers Street. Ground floors were devoted to shops, upper floors to living quarters. In the twelfth and early thirteenth centuries, four stout posts sufficed to frame the house, with horizontal members tenoned into mortises in the posts, to which the walls were secured by wooden pegs;[72] in the late thirteenth and fourteenth centuries, such houses were gradually replaced by timber-framed buildings with stone foundations. The poorer the neighborhood the narrower the house: a street excavated in Winchester revealed a row of houses measuring fifteen by fifteen feet.[73] Shops were essentially stalls, with fronts that opened for business by letting down a hinged section on which merchandise could be displayed. Windows were covered with oiled paper or parchment, or cloth coated with a compound of white wax and resin.[74] The extent to which craftsmen (and craftswomen, wives typically assisting husbands) dominated the life of the city is indicated by data of the late thirteenth century showing that out of 50,000 inhabitants of Bologna no fewer than 36,000 were members of guilds or relatives of members.[75]

Cities were expensive to live in because of the need to import everything from outside, often including water to supplement fountain, well, or cistern, with a profit to the middleman added to the cost of transport. The larger the city the higher the cost of living. When the bishop of Bath and Wells moved to London in 1338, his household accounts showed prices rising by 33 to 100 percent. A pig that cost the bishop two shillings in Somerset cost him three in London, and the prices of candles, oats, and ale nearly doubled. Besides being more expensive, cities were less healthful than the countryside. Archaeologists report large numbers of intestinal parasites in the cesspits, while by the fifteenth century the endemic Black Death had come to be known as primarily an urban malady.[76]

Of all the medieval cities, those most clearly foreshadowing the future were the great cloth towns of Flanders and Italy, where in place of the many specialized crafts of the smaller cities the dominant textile industry created harsh class differences. The houses of the rich drapers like Jehan Boinebroke clustered in Europe's first *beau quartier* residential districts, while the warrens of tenements that housed the families of the weavers formed the first proletarian slums.

The Gothic Engineer: Villard de Honnecourt

Dominating the skylines of cities across medieval Europe now rose what W. H. Auden calls the "Plainly Visible Churches: / Men camped like tourists under their tremendous shadows."[77] Nothing like the Christian cathedrals had ever been seen in cities before, yet they became at once familiar and even convenient additions, often used for secular purposes and even serving as cradle for the medieval mystery plays that grew into the modern Western theater.

The master mason who directed construction of these majestic and useful monuments came to be regarded with an esteem that belied his typically common origins. The most telling sign of his standing in the Christian community is the striking image in medieval art that depicted God as a master mason, holding scales, carpenter's square, and compasses, the tools of his trade. Master masons themselves were typically represented wearing cap and gown, like university masters.[78] The men who inspired such respect also inspired envy. Gervase, the monk who described the rebuilding of Canterbury Cathedral, admired William of Sens for his competence but could not forbear expressing some question about the hubris of such men. In the accident that disabled William, he noted that "no other person than himself was in the least injured. Against the master only was this vengeance of God or spite of the devil directed." Jacques de Vitry, Paris preacher known for his biting social criticism, described the master mason on the job: "He orders his men about but rarely or never lends his own hand. Pointing his walking stick, he directs, 'Cut here,' or 'Cut there' [and is promptly obeyed]."[79] Others echoed the stricture. Nicolas de Biard complained, "Masters of masons . . . do no labor, and yet they receive a higher fee" than the ordinary stonemasons.[80]

King confers with master mason. [From Matthew Paris, Life of Offa, British Library, Ms. Cotton Nero, D I, f. 23v.]

Such men did not limit their activities to cathedral building but directed construction of all types—cloth halls, warehouses, hospitals, markets, town walls, and castles. In their military roles they were commonly called engineers, or enginers, as in Hamlet's reference to "the enginer hoist with his own petard [explosive device]."[81]

The names of hundreds of masters are known, along with those of many other cathedral workers—sculptors, carpenters, painters, lead workers, glaziers, draftsmen. Their signatures often remain inscribed on their work. Some, such as Pierre Montreuil, builder of Sainte-Chapelle in Paris, were buried in their churches. But although we know their names, we know less than we could wish of their methods. They were not deficient in general education—not only could they read and write but they had some command of geometry and arithmetic. What they lacked was engineering theory, in place of which they employed their own experience, that of colleagues, and rule of thumb.[82] Instead of the modern engineer's blueprints, computer models, and other planning tools, they had the "tracing floor," a smoothed area on which details of arches, piers, and windows could be drawn full size. They used sketches as well as written and oral communication, and guided their stonemasons at the quarry with molds—models in wood or plaster fashioned by a carpenter.[83] A cord attached to a fixed point was used to mark out large arcs.[84] To achieve precision in floor plans, they employed compasses and the L-shaped carpenter's square, the latter often made from the thoroughly seasoned wood of used wine barrels.[85] If the resulting cathedrals were intensely spiritual, they were also "intensely geometrical" (Arnold Pacey).[86]

While the stone was being prepared at the quarry, the timber falsework to support arches and vaults during construction was erected. Then began the heavy and exacting work of lifting the stones one by one into place by windlass-power hoist and the "Great Wheel," powered by treadmill, mounted on the roof

Building the Tower of Babel, c. 1250: a treadwheel used to lift stone supplements hod carriers and stretcher bearers. [The Pierpont Morgan Library, M. 638, f. 3.]

Building the Tower of Babel, c. 1430: crank-style windlass with flywheel is used to lift stone. In the foreground, masons work with compass, T-square, hammer, and chisel. [British Library, Bedford Hours, Ms. Add. 18850, f. 17v.]

beams above the vault, where it sometimes remained permanently.[87] As the structure rose skyward, scaffolding was built against it, usually in the shape of an inclined ramp supported on poles. Passageways (vices) and stairways were built into the fabric of many cathedrals, as at Chartres, providing stable enclosed access to construction points.

In the earlier cathedrals nearly every detail of construction was personally supervised by the master mason. In the thirteenth century, as Jacques de Vitry's description indicates, the master took on more of the character of general of an army, with subordinate officers and a labor force comparable in size to most real medieval armies. Master James of St. George, building Beaumaris Castle in the closing years of the thirteenth century, had under his orders a force of 400 masons, 2,000 laborers, 200 quarrymen, 30 smiths and carpenters, and operating equipment that included 100 carts, 60 wagons, and 30 boats to carry material to the building site.[88] Sometimes tasks were subcontracted; at Windsor Castle in 1362–1368, John Martyn, John Welot, and Hugh Kympton all contracted to build vaults and the corresponding wall sections while the overall project remained under the direction of two masters, John de Sponlee and William Synford. Some famous masters engaged in more than one project at a time, requiring the employment of assistant masters at the sites.[89]

The master mason's army was divided into two cohorts, one cutting stone at the quarry, the other erecting it at the site. Stone was cut with saw and bush hammer and given final shap-

Masons use a windlass with radiating spokes, plumb lines, levels, axes, and adzes. [From Matthew Paris, Life of Offa, Trinity College Library, Dublin, Ms. 177, f. 60v.]

ing with mallet and chisel, to reduce to a minimum the weight to be transported. Volunteer labor, paid with indulgences, was sometimes enlisted for the transport, but horses with the collar harness were the main reliance. All the important labor, skilled and unskilled, was hired, the unskilled recruited locally, the skilled—masons, glaziers, lead workers—nearly all itinerant, in latter-day terminology "boomers," who took their well-paid expertise from construction site to construction site, often crossing national boundaries.[90]

Masters exchanged information, enriching each other's practical backgrounds. At least one master composed a manual designed to supply such material to others. The large-format notebook of Villard de Honnecourt has survived (at least in great part) to provide data on Gothic engineering, including plans, elevations, sketches of building machinery, and other

A page from Villard de Honnecourt's sketchbook. The caption reads, "Here begins the method of representation as taught by the art of geometry, to facilitate work." [From The Sketchbook of Villard de Honnecourt, ed. by Theodore Bowie, Indiana University Press.]

details. Villard was a mason from Picardy who composed his book while working and traveling to sites in Rheims, Chartres, Laon, Meaux, Lausanne, and even Hungary. The text is in French, but internal evidence shows that its author was also literate in Latin.[91]

The book's purpose, as Villard explained it, was to teach his successors not only how to use wood and stone in construction but how to apply rules of geometry to portraiture and design. He used squares, triangles, and other figures to aid in drawing human beings and animals, in taking the elevation of a structure from the plan, in positioning a building optimally in a given space, and in calculating the height of a construction, the width of a stream, and the exact center of a site. Writing of what "the art of geometry commands and teaches," Villard expressed "a philosophic conviction suggestive of Platonism"

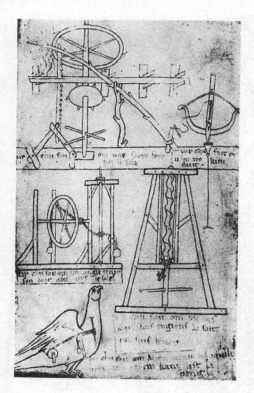

Mechanical saw, top left; middle left, device that may have been a primitive escapement; right, screw jack. [From The Sketchbook of Villard de Honnecourt, ed. by Theodore Bowie, Indiana University Press.]

(Arnold Pacey).[92] Many of the tricks Villard taught for transferring drawings from parchment to stone block, wall, or glassmaker's table later became, under the influence of the masons' guilds, trade secrets that guildsmen were forbidden to divulge to outsiders.[93]

Among sketches of hoisting machinery, treadmills, windlasses, and other devices available to the thirteenth-century master mason, Villard pictured a waterpowered saw, whose downstroke was effected by the turn of the waterwheel and which was returned to its original position by a spring in the form of a sapling bent back by the downstroke, the first representation of two motions applied automatically to a mechanism.[94]

Similar sapling springs were used to reverse the motion of a lathe, to which a refinement was added in the form of a foot treadle. A Chartres window shows a double-treadle lathe in which the strap passes through a pulley fastened to the ceiling.[95] Villard also depicted a screw jack, expressing astonishment at the power of this simple lifting device, with the implication that it was of recent provenance.

The master mason's rule-of-thumb methods led to many mistakes (as was still true of his nineteenth-century successors). Late in the construction of Chartres Cathedral, additional flying buttresses, copied from those at Notre Dame de Paris, were added. The highly sloped buttresses of Bourges Cathedral, built

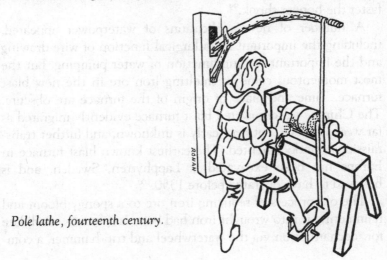

Pole lathe, fourteenth century.

at the same time, probably also reflect the experience gained at Notre Dame with the effect of wind on tall structures.[96] Competition in height akin to that in twentieth-century American skyscraper construction led to a record spire at Strasbourg of 468 feet, equivalent to the height of a modern forty-six-story building, but also to the collapse of the nave at Beauvais in 1284, which put a damper on the competition.

Metallurgy: The Waterpowered Blast Furnace

If the cathedral was the aesthetic marvel of the Middle Ages, a less prepossessing structure was, in the opinion of R. J. Forbes, "the greatest technical achievement" of the period.[97] In the medieval invention of the blast furnace, the waterwheel once more played a central role.

The spread of waterpower from tributaries and small rivers to the larger rivers was made possible by the construction of dams and millraces, and was signaled in the documentary record by the marked increase after 1300 in laws and lawsuits involving navigation rights versus power rights.[98] The vertical waterwheel acquired new accessories, such as the mechanical governor that helped grind the grain and sift the flour at Clairvaux and elsewhere: a square segment of the millstone axle acted as a cam, catching against a projection on the hopper, causing it to shake and discharge its flour. The faster the waterwheel turned, the faster the hopper shook.[99]

A number of new applications of waterpower appeared, including the important metallurgical function of wire drawing and the important mining function of water pumping, but the most momentous came in smelting iron ore in the new blast furnace. Time and place of origin of the furnace are obscure. The Chinese waterpowered blast furnace evidently migrated as far west as Persia, but how early is unknown, and further transmission is undocumented. The earliest known blast furnace in Europe has been excavated at Lapphyttan, Sweden, and is believed to have operated before 1350.[100]

The old process of reducing iron ore to a spongy bloom and hammering it into wrought iron had been an obvious candidate for mechanization via the waterwheel and trip-hammer, a combination in wide use by the fourteenth century. The water-wheel was now enlisted to pump pairs of bellows several feet in diameter, mounted in tandem and blowing alternately through a common tuyere, increasing the draft and decisively raising the temperature in the furnace. The draft was also increased by the furnace's new form. What had once been little more than a pit and a stubby chimney had gradually risen into a novel shape: a tall masonry structure square in plan, mounted over a crucible (firebox) built on a flat stone hearth. The chimney was made up of two vertical pieces, a short lower one shaped like an inverted truncated pyramid (resembling a grain mill hopper in profile), topped by a tall right-side-up truncated pyramid.[101] The whole structure rose eighteen or twenty feet above ground, though the hearth within was no more than a foot square. By 1400 blast furnaces were operating (in addition to Sweden) in Styria (Austria), the Rhine valley, and the neighborhood of Liège (Belgium).[102]

The stronger blast of air in the new furnaces heated the ore to a point where carbon uptake became very rapid, producing an alloy of about 4 percent carbon and 96 percent iron. This metal had a much lower melting point than pure iron (about $1,100°$ C as against about $1,530°$ C), making possible the casting of molten iron. Almost at a stroke the blast furnace carried the ancient handicraft of iron making into the industrial age. Cast iron became the sought-after intermediate product of an entirely new two-stage process.

A waterpowered blast furnace could run continuously, for weeks or months at a time. The sand and clay containing the iron ore were mixed with a limestone flux to form the furnace's charge, which was layered alternately with charcoal. As the mass heated, the sand, clay, and limestone formed a slag that floated on top of the heavier molten iron. The slag was removed periodically from an opening near the top of the furnace, the iron run off through another at the bottom.[103] In early blast furnaces, the iron ran into a bed of sand to cool in successive batches, but the quantities that could be produced brought about an expansion of the sand bed into a system given a picturesque medieval nomenclature. Starting in a canal called the "runner," the molten metal flowed into several large, shallow depressions. The image of the depressions reminded smiths of a sow with suckling pigs, and the term "pig iron" was born.

The cooled pig, weighing a couple of hundred pounds, was transported to a secondary furnace called a "finery," a charcoal-fired hearth equipped with two air blasts, one to supply draft for the fire and another to play on the iron as it heated, its oxygen combining with the carbon in the metal and blowing off in smoke, leaving pure (wrought) iron. These air blasts were also soon powered by waterwheel and continued to be on into the nineteenth century, when no one could remember why the iron chunks were called pigs.[104]

The new system produced much more iron with much less labor, reducing cost and multiplying applications. It did not bring an immediate shift to the casting of iron implements. The smith continued to work at his forge, with either pig or bloomery iron, first shearing a piece of roughly the proper size with chisel and hammer, then reheating and hammering into shape as blade for sickle, scythe, ax, adze, or mattock, as fire tong, hinge, tip for spade, wool comb, axle part, or the universal cauldron, used for cooking, brewing, and bathing the baby.[105]

A product of the smith: the universal cauldron, used for cooking, brewing, and bathing the baby. [British Library, Ms. Cotton Claudius B IV, f. 28.]

The fourteenth-century smith still commanded respect, but he had become less of a mysterious specialist in aristocratic arms and armor and more of a homely and familiar figure in the community, valued as a craftsman, but not always welcome as a neighbor. A contemporary poem entitled "A Complaint Against the Blacksmiths" gives a picture of the forge in the alliterative style of *Piers Plowman*:

> The crooked codgers cry after: Coal! Coal!
> And blow their bellows till their brains are all
>   bursting.
> Huff! Puff! says the one, Haff! Paff! says the other.
> They spit and they sprawl and they tell many tales.
> They gnaw and they gnash and they groan all
>   together
> And hold themselves hot with their hard hammers.
> Of a bull's hide are built their bellies' aprons,
> Their shanks are sheathed against flickering flames.
> Heavy hammers they have that are hard to handle.
> Stark strokes they strike on a stock of steel.[106]

In 1397 in London, smiths were being invited to leave neighborhoods because of "the great nuisance, noise, and alarm experienced in divers ways by neighbors around their dwellings." A spin-off branch of the trade was found even more objectionable. The spurriers (spur makers) were reputed to "wander about all day without working," getting drunk and "blow[ing] up their fires so vigorously" at night that they blazed, "to the great peril of themselves and the whole neighborhood." In 1377 the neighbors of a London armorer named Stephen atte Fryth lodged a formal complaint against him, alleging that "the blows of the sledge-hammer when the great pieces of iron . . . are being wrought into . . . armor, shake the stone and earthen party walls of the plaintiffs' house so that they are in danger of collapsing, and disturb the rest of the plaintiffs and their servants, day and night, and spoil the wine and ale in their cellar, and the stench of the smoke from the sea-coal used in the forge penetrates their hall and chambers."[107]

"The Most Pernicious Arts": Firearms from China

The blast furnace arrived in the West just as a new use for metal appeared, quite suddenly but with little fanfare: firearms. In China, gunpowder weapons had matured over some four centuries, from alchemists' experiments with explosive mixtures to primitive guns embodying three basic features: a metal barrel, a dependable explosive, and a projectile efficiently fitted to the bore.

While the firearms evolution proceeded in China, Europe continued to tinker with the crossbow. The English longbow, actually Welsh in origin, played so conspicuous a role in the English victories of Crécy (1346) and Poitiers (1356) that debate over the rival merits of the two bows has continued into the twentieth century. Despite the longbow's more rapid rate of fi·e, the decisive evidence in favor of the crossbow seems to be the failure of the longbow to diffuse on the Continent and the fact that, despite Crécy, Poitiers, and Agincourt (1415), the French won the Hundred Years War. In any case, it was the crossbow that was susceptible of technical improvement, which it received in two directions. The old wood, bone, and composition materials were replaced, from about 1370, by steel. The resulting bow had an extreme range of 400 to 450 yards and required a more powerful cocking mechanism, three different forms of which were invented. The "goat's foot" was a long lever atop the stock, the cranequin a ratchet device moved by a horizontal crank, and the windlass a winch powered by a small double crank.[108]

More effective bows and greater availability of iron brought on a defensive reaction: a steady increase in the use of plate armor. The mature coat of mail, or hauberk, fashioned of interlinked iron rings, remained through the first half of the fourteenth century the fundamental protection of the torso, with plates added to cover arms and legs. The articulation needed to permit freedom of movement was achieved mainly through "lames," overlapping leaves pinned by rivets fixed to one piece and sliding along a slot in its neighbor. By the fifteenth century the knight "in full armor" was a familiar battlefield sight.[109]

Other innovations were in the air. In 1335 Guido da Vigevano, royal physician and astrologer at the French court, proposed what amounted to history's first tank, an armored wagon powered by a windmill mounted on top. In a more practical vein, Guido also suggested pontoon bridges and assault towers fabricated in small interchangeable sections that could be transported by pack animal and assembled in the field (his patron, Philip VI of France, was contemplating a Crusade). Guido's treatise has been called (by Bertrand Gille) a milestone between the notebook of Villard de Honnecourt and the great engineering sketchbooks of the fifteenth century.[110]

Guido made no mention of firearms, which, however, had by this time made their unobtrusive entry on the stage. The first European mention of gunpowder occurs in 1268 in the writings

Crossbow confronts longbow at the Battle of Crécy, 1346. From the Chroniques of Froissart. *[Bibliothèque de l'Arsenal, Ms. 5187, f. 135v.]*

of the English Franciscan friar Roger Bacon, in a passage that Joseph Needham believes to be a description of Chinese firecrackers:

> We have an example of these things . . . in that children's toy which is made in many parts of the world: i.e., a device no bigger than one's thumb. From the violence of that salt called saltpeter together with sulfur and willow charcoal, combined into a powder, so horrible a sound is made by the bursting of a thing so small, no more than a bit of parchment containing it, that we find the ear assaulted by a noise exceeding the roar of strong thunder, and a flash brighter than the most brilliant lightning. Especially if one is taken unawares, this terrible flash is very alarming. If an instrument of large size were used, no one could withstand the noise and blinding light, and if the instrument ere made of solid material, the violence of the explosion would be much greater.[111]

How did Roger Bacon learn about Chinese fireworks? A possible explanation lies in the eastward journey of William of Rubruck a few years earlier. One of a number of European missionaries to visit China in the mid–thirteenth century, William was a fellow Franciscan and personal acquaintance of Roger. Needham speculates that William described Chinese firecrackers to his friend, or even brought some back with him as a curiosity.[112]

The employment of volatile mixtures in war had been familiar to both Europeans and Arabs ever since Greek fire was first used in the seventh century. By the same token, so was their discharge from a metal tube. But the use of such a mixture as a missile propellant was something new. Suddenly in the fourteenth century, niter, the sodium or potassium salt of nitric acid, also known as saltpeter, became the object of systematic collection from European barns, stables, and pigsties, to be mixed with sulfur and charcoal and ignited in metal tubes to propel missiles.

How this development came about remains a tantalizing mystery. Needham proposes three separate channels of communication from China: First, knowledge of gunpowder chemistry via missionaries like William of Rubruck or other European travelers. Second, knowledge arriving via the Arabs (a Spanish Muslim scientist referred to saltpeter as "Chinese snow") of bombs, rockets, and a weapon called the fire-lance, a bamboo, wood, or metal tube that spouted a mixture of pellets, pottery shards, and toxic chemicals in a stream that lasted some minutes. Third, by about 1300, knowledge of metal-barreled guns, possibly conveyed overland through Russia. That the Chinese were making gun barrels as early as 1300 is known from archaeological finds.[113]

Little that is conclusive can be adduced from the evidence. Except for the passage in Roger Bacon, no trace appears in either European or Islamic records of the kind of fumbling experimental steps by which China progressed to gunpowder weapons. Instead, a Florentine document of 1326 describes the city authorities' acquisition of metal cannon and iron shot in language indicating that the items were by then common-

Franciscan friar Roger Bacon. [Bodleian Library, Ms. Bodl. 211, p. 5.]

place.[114] In light of the Florentine document, Carlo Cipolla believes the "invention" of cannon to go back to the late thirteenth century.[115] The earliest documented use of cannon in Europe was by two German knights at the siege of Cividale in northern Italy in 1331. Edward III brought at least twenty guns and large quantities of sulfur and saltpeter to the siege of Calais in 1346.[116] Noteworthy is the fact that whatever the history of diffusion from China, Europeans had at this point not only overtaken the Chinese in firearms but surpassed them, since guns large enough to call cannon had not yet been manufactured in China, where cannon first appeared in the anti-Mongol revolution of 1356–1368.[117]

In short, the priority of the invention of firearms is incontestably Chinese, and a high degree of probability exists, that most or all of the necessary knowledge was received by Europe from China. Yet some independent European contribution was involved, and Europe displayed an enthusiasm for the new weaponry that contrasts with Chinese indifference. Writing in the 1350s, Petrarch noted, "These instruments were a few years ago very rare . . . but now they are become as common and familiar as any other kind of arms. So quick and ingenious are the minds of men in learning the most pernicious arts."[118]

Early European cannon were made of copper, brass, or bronze, but a technique was soon devised for using the cheaper iron produced by the blast furnace as a practical gun material: the smith welded a cylinder of iron rods around a clay core to form a barrel, which he strengthened by shrinking iron bands around it. The core was then dug out. Cannonballs were first made of lead or iron, then of cheaper stone, which the stonecutters fashioned with the aid of a "patron" or template of wood, parchment, or paper. But when it became possible to cast cannonballs of iron, stone lost its advantage in price. Iron balls may also have provided a better fit to gun bores. By 1418 the city of Ghent was ordering 7,200 cast-iron cannonballs.[119] Gunpowder was mixed in the field by the cannoneers, who were usually the same smiths who fabricated the cannon. Opinion varied on the proportions of saltpeter, sulfur, and carbon (charcoal), but medieval saltpeter content generally ran close to the 75 percent used for modern black powder.[120] Premature explosions were common.

The first European handguns, which appeared at the end of the fourteenth century, suffered from other deficiencies. The gunner heated a wire red-hot, then had to aim his weapon while inserting the hot wire into a touchhole on the top of the barrel. A two-man version was easier to use—one man balancing the gun on his shoulder like a World War II bazooka while his mate applied the wire—but accidents were frequent.

At the point when the first half of the Hundred Years War was terminated by a truce (1396), the new weapon had yet to prove its value. Despite greater range and accuracy and a more

Cannon on shipboard, with gun ports, 1482. [Bibliothèque Nationale, Ms. fr. 38, f. 157v.]

rapid rate of fire, it only slowly displaced the trebuchet (which threw a heavier missile). Unlike a trebuchet, a cannon could not be assembled in the field, nor was hauling it long distances easy. Two wagons in line provided a form of articulation, but the contraption often overturned. Arrived in the field, the gun had to be set up on a frame or trestle for firing, generally with mediocre effect.[121] At the battle of Aljubarrota in 1385, the Castilians employed sixteen "great bombards," but the Portuguese, who had no cannon at all, won the battle.[122]

Early employment of gunpowder weapons at sea brought equally unimpressive results and turned up some fresh problems. Galleys, with their low freeboard, proved poor platforms for artillery, and on the decks and castles of deeper-hulled vessels cannon created top heaviness and instability in foul weather. The solution, a gun deck with gun ports pierced in the hull, was not found until the following century, when it introduced a whole new mode of naval warfare.

"A Wonderful Clock"

In the advance of Europe to the forefront of world technology, the emergence of the mechanical weight-driven clock in the second quarter of the fourteenth century has been widely regarded as a decisive moment. Donald Hill calls it "one of the main foundations for the development of machine technology in subsequent centuries,"[123] and D. S. L. Cardwell describes it as "perhaps the greatest single human invention since the wheel."[124]

At one time it was believed that the Western mechanical clock came into being in response to the monasteries' need for better timekeeping devices to govern their system of canonical hours. But the clepsydra adequately satisfied monastic needs, and in the early evolution of the clock, timekeeping was actually a secondary consideration. In Europe, as previously in Asia, clockwork developed out of the demand for precision instruments to aid in tracking stars and planets. The demand came from the astrologers, whose science was by now an established part of medical practice. Two of the clock's ancestors were the astrolabe, in its improved Islamic form, and the equatorium, another Muslim instrument, used to calculate positions of the planets on the basis of Ptolemy's system. The earliest Western "clocks," such as a famous one built by Richard of Wallingford in about 1320, have been described as "powered astronomical models," "artificial universes," and "pre-clocks."[125]

Weights as driving mechanisms had long been known, and gearing was by now thoroughly familiar to Europe's metal craftsmen. What was necessary to translate the gravitational pull into controlled motion was a means of governing the descent of the weight, whose natural tendency was to fall at an accelerating pace.[126] The complex escapement of Su Sung, designed for a water-driven wheel, was never known in Europe. Villard de Honnecourt's notebook contains a sketch in which the statue of an angel is made to point continuously toward the sun by a wheel whose spokes strike a taut rope stretched by two weights: an escapement, but one so crude (and crudely represented) that it has only slowly been recognized as such and probably had no influence on the invention of the true clockwork escapement.[127]

The fact that Latin and the western European languages had no special terms to distinguish mechanical from water clocks has helped to obscure the story for modern historians. Abbott Payson Usher collected twenty references to clocks dating from between 1284 and 1335, all of which upon investigation turned out to be water clocks.[128] Both Dante's *Inferno* (1308–1321) and the late-thirteenth-century *Roman de la rose* contain literary references over which debate remains inconclusive. The origin of the European escapement is almost surely lost forever, but consensus today places it in the second half of the thirteenth century,[129] and its emergence in the historical record signals its provenance as northern Italy. Joseph Needham believes in the possibility of stimulus diffusion of the idea of an escapement in the form of travelers' tales from China, but this hypothesis seems farfetched. As Carlo Cipolla says, "The Chinese escapement . . . had nothing in common with the European verge-and-foliot device."[130] What is certain is that the verge-and-foliot (or crown wheel and foliot) escapement is one of the most elegant solutions ever devised to a problem in mechanical engineering.

The essential parts of a verge-and-foliot escapement were the crown wheel, with triangular teeth set perpendicularly around its edge (like the points on a crown); the verge or rod, standing close to it, with two projections (pallets) perpendicular to each other, so placed as to engage with the crown wheel at its top and bottom; and the foliot, a crossbar balanced at the top of the verge, with weights at each end. As the weight-driven crown wheel turned, one of its teeth caught the upper pallet of the verge, which held it momentarily and then released it, giving a swing to the foliot with its weights. This caused the other pallet to engage the wheel, swinging the foliot in the opposite direction. Thus the wheel's motion was alternately arrested by the two pallets of the verge, and as the foliot swung back and forth, the wheel turned a click at a time. To regulate the clock, the speed of the mechanism could be increased or decreased by moving the weights on the arms of the foliot. One of the insights of the unknown inventor was the

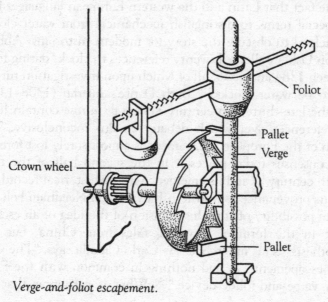

Verge-and-foliot escapement.

fact that the top and bottom of a revolving wheel are moving in opposite directions.[131]

The first mechanical clocks were huge iron-framed mechanisms, fabricated by blacksmiths and installed in towers; St. Eustorgio in Milan had one as early as 1309. They had no face or hands and did not strike the hours, but merely sounded an alarm which alerted the ringer to pull the bell rope. In 1335 the first clock that struck automatically, to the astonished admiration of the citizens, was also erected in Milan, in the tower of the Visconti palace chapel. "There is a wonderful clock with a very large clapper," wrote a contemporary, "which strikes a bell twenty-four times according to the twenty-four hours of the day and night, and thus at the first hour of the night gives one sound, at the second two strokes . . . and so distinguishes one hour from another which is of the greatest use to men of every degree."[132]

L'Horloge de Sapience (*The Clock of Wisdom*), c. 1450. At left, a clock with an hour hand and twenty-four-hour face; an astrolabe hangs below it. The table on the right bears a clock that may be the first evidence of a spring-driven timekeeper. [Bibliothèque Royale, Brussels, Ms. IV, III, f. 13v.]

The earliest known makers of real (timekeeping) mechanical clocks are Jacopo di Dondi and his son Giovanni. In 1344 Jacopo created a clock for the entrance tower of the Carrara palace at Padua, which besides automatically indicating "the intervals of four-and-twenty hours by day and night" showed the phases of the moon and other astronomical features. A clock built by Giovanni di Dondi for the castle of Pavia, installed in 1364, has been described as "a true mechanical clock," equipped with weight drive, verge-and-foliot escapement, seven dials with gear wheels and linkages to show astronomical motions, a fully automated calendar showing the holy days, and, almost as an afterthought, a small dial for telling time.[133]

Once the verge-and-foliot escapement became known, blacksmiths in cities all over Europe began turning out clocks. By 1370 at least thirty had been installed, all with timekeeping an inconspicuous and only moderately successful function.[134] The astronomical garnishment, on the other hand, quickly came to serve an aesthetic as well as a scientific function. The city clock became a source of civic pride, "a marvel, an ornament, a plaything . . . a part of the municipal adornment, more a prestige item than a utilitarian device" (Jacques Le Goff).[135] The enormous clock built at Strasbourg in 1354 included a moving calendar; an astrolabe whose pointers indicated the movements of the sun, moon, and planets; a statue of the Virgin before whom every noontime the Magi bowed while the carillon played a tune; and, atop the whole, a large cock that opened its beak, crowed, and flapped its wings.[136]

The attention given by the clockmaking smiths to such ornate details may have detracted from that given to precision instrumentation. Most medieval clocks gained or lost many minutes in twenty-four hours. At first nobody cared very much. Contemporary requirements for accuracy were liberal. Though astronomers had by now subdivided the day's hours into sixty-second minutes, a system borrowed from ancient Babylon, medieval people had long been accustomed to variable winter and summer hours, to fit the daylight available for work.

It was the new clocks, with their noisy officiousness, that gradually imposed the system of equal hours, causing people to begin timing activities that no one had thought of timing before. In the cloth-making towns of Flanders, the clocks struck the working hours of the textile workers. Forthwith, "the communal clock [became] an instrument of economic, social, and political domination wielded by the merchants who ran the commune" (Le Goff).[137] In Paris in 1370, Charles V ordered all the bells of the city to keep time with the clock in the Palais-Royal as it rang the hours and quarter hours, regimenting the city into a uniform time frame. Uniform, but not notably reliable, as a Parisian verse observed:

> L'horloge du palais
> Elle vas comme il lui plait.[138]

(The palace clock/It goes as it pleases.) The uniformity was local rather than national. Each city set its own zero hour— sometimes noon, sometimes midnight, but more often sunrise or sunset, creating a confusion that continued to baffle travelers into the fifteenth and sixteenth centuries.

The first household clocks appeared shortly before 1400. In contrast to the big tower clocks made by blacksmiths, the smaller versions had faces, hour hands, and later minute hands, and were the work of goldsmiths and silversmiths.[139]

One of the most significant things about medieval clockwork is simply that these were the very first machines made entirely of metal; all preceding machinery had been mainly wooden. The metal smiths' tradition of precision work, here established, lasted all the way into the eighteenth century, when it gave them a key role in fabricating and operating the textile machinery of the Industrial Revolution.

Meanwhile, the mechanical clock, invented as an almost incidental component of a mechanism designed to serve the needs of the pseudoscience of astrology, rapidly acquired its own significance. Once people could time their activities, they subordinated them to time, working and living by the hour, in a new rhythm that continues to this day.

Wheels of Travel, Wheels of Commerce

As the Commercial Revolution increased the strain on road surfaces, it also heightened the influence of the merchants who paid the tolls and taxes. For the first time serious effort went into road maintenance. Old roads were repaired and new ones built, employing the established technique of cobbles or broken stone on a foundation of loose sand. Not as strong and rigid as the Roman road, the medieval road was easier to maintain and on the whole better suited to vehicular traffic. In some locations, mainly within cities, mortared paving blocks were used.[140] All over northwest Europe the road paver became a familiar sight, in France and Flanders sitting on a four-legged stool and moving forward as he worked, in Germany sitting on a one-legged stool and moving backward.[141]

Where grades were excessive for wagons, as so often in Switzerland, standby draft animals were stationed to be added to teams as needed. In 1237 a new road and "daring bridges" (Robert Lopez) opened the St. Gothard Pass to pack animals. One narrow stone arch over the rapids of the Reuss was called the Pont Ecumant (Foaming Bridge) because of the spray that perpetually drenched drivers and animals. In the following century the Swiss opened the first Alpine road capable of accommodating wheeled traffic, over Monte Settimo.[142]

Elsewhere, the bridge-building boom of the eleventh and twelfth centuries peaked in the thirteenth.[143] In the Ile-de-France, eight new bridges were built in the eleventh century, seventeen in the twelfth, and thirty-four in the thirteenth.[144] As commerce furnished an ever larger element in the traffic, the towns increasingly assumed responsibility for bridge construction and maintenance. The old Roman concept of bridges as public works revived, together with the idea of financing them by taxation. St. Bénézet's Pont d'Avignon passed from the aegis of the Bridge Brothers to that of the communal government. The tradition of private support did not die out, however. In the fifteenth century, donations were still being received, such as that following a disastrous flood of the Loire from "a person who had great love and affection for the bridge [at Orléans] and its rebuilding."[145]

Some new construction techniques appeared. One of Villard de Honnecourt's sketches shows a machine for sawing off the tops of bridge piles.[146] The Roman cofferdam came back into use; saplings were driven into the riverbed to form an enclosure which was pumped out, and the piles driven for the pier foundation, with a core of rubble or stamped-down clay and masonry blocks laid on top. By this rough-and-ready method, several bridges of record-breaking length were built: the 54-meter (175-foot) single arch over the Allier at Vieille-Brioude, in southern France, built in the 1340s (and lasting until 1822)[147] and in the 1370s the even longer arch over the Adda at Trezzo in northern Italy, at 72 meters (236 feet) the world's longest single-arch span until the eighteenth century. The Karlsbrücke at Prague, begun by the Emperor Karl IV to bridge the broad Moldau, took forever to finish, owing to the Hussite wars and other interruptions, but when finally completed in 1503 it was, at nearly 600 meters (1,970 feet), the world's longest stone-arch bridge.[148]

Regular contracts were now awarded to masons and to the

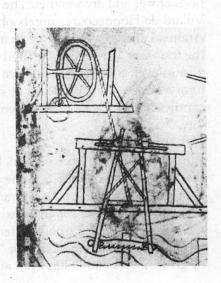

The caption reads: "By this means, one can cut off the tops of piles under water so as to set a pier on them." [From The Notebook of Villard de Honnecourt, ed. by Theodore Bowie, Indiana University Press.]

carpenters who built the falsework to support the arches during construction. For some bridges, especially those near cities, periodic inspections were carried out by teams of masons, carpenters, and communal officials. Innovations in design were few, though a fine model of the segmental arch bridge appeared in Florence in 1345, in what came to be known as the Ponte Vecchio (Old Bridge). Whether Taddeo Gaddi, its architect, was acquainted with the Pont-St.-Esprit in France is unknown; his segmental arch was a novelty in Italy. The reduction in scour effected by the narrower piers of the segmental form helped the bridge, crowded with shops, houses, and tourists, to survive Arno floods to the present day.

Far more numerous than the stone bridges were those built of timber, but few traces of these remain. Two survivors are the famous Kapellbrücke and Spreuerbrücke of Lucerne, Switzerland, known chiefly for their mural galleries but of technical importance for their partial truss construction. The truss design, based on the structural strength of the triangle, had a long history as a roof support but was only tardily exploited as a bridge form. The function of the picturesque roof and siding was to protect the structural members of the truss from alternations of wet and dry weather. The inclusion of a kind of truss in Villard de Honnecourt's notebook, and of several variations in Andrea Palladio's *Treatise on Architecture* of 1570, suggests that the form was widely used for short crossings. Villard also depicts a cantilever, a balanced structure usually employed in pairs to form a bridge. Long produced in stone in China and India, the cantilever was not adopted in Europe until the nineteenth century.[149]

While the total number of vehicles of all types multiplied on the roads, shifts took place in the proportions of the categories. Four-wheeled wagons, long outnumbered by two-wheeled carts, became much more common, while at the upper end of the social scale the first carriages offering a degree of comfort were introduced. *Chariots branlants*, or "rocking carriages," were used by great nobles and ladies from at least the 1370s. Until then, in the words of Marjorie Boyer, "the chassis of a lady's personal *char* was essentially no different from the chariot in which her baggage was transported," the carriage body resting directly on the axles and transmitting every bump in the road to its occupants. The "rocking carriage," employing chains hung from posts run transversely under the body, somewhat ameliorated the jolting. An illustration from a Zurich manuscript of the mid–fourteenth century shows what seems to be the longitudinal suspension of a carriage body from leather straps, but such carriages did not arrive in numbers in western Europe until the following century. Their origin was Hungary, where the town of Kocs (hence "coach," *coche*, *Kutsche*) became famous for its lightweight, one-horse, leather-suspended passenger vehicles.[150]

Eventually wagons were also improved by suspension, although the change came only slowly, and to the end of the Middle Ages merchandise was damaged and the wagons themselves were jarred to pieces by the unrelieved shocks of ruts and potholes. Protection of merchandise from the weather, however, was effected by the *longa caretta*, twelfth-century ancestor of the Conestoga wagon.[151]

The advance to four-wheeled wagons was assisted by the movable forecarriage, which appeared before the end of the fourteenth century, greatly reducing the turning radius, but once more general adoption was slow.[152] Wheels were provided with iron tires in the form of a number of small plates, clumsily nailed on. The technique of shrinking heated bands onto wheels was not invented until the sixteenth century.

Despite shortcomings, wagons were stronger and more durable, and animal harness more efficient than in previous centuries. For carriages, the breast harness was favored, in the shape of a long leather strap passing completely around the animal horizontally, forming a strap across the chest to pull against and a strap across the rump to hold back the weight when descending a hill.[153]

By the high Middle Ages, land transport was significantly cheaper. Wagons could carry goods twenty-two to thirty-five kilometers (fourteen to twenty-two miles) a day in level country, adding a transportation cost per eighty kilometers (fifty miles) traveled that for wool amounted to only about 1.5 percent, for grain about 15 percent.[154] Pack animals, which still carried most of the freight, achieved even better speed.[155] In the summer of 1375, William de Percelay carried sacks of silver pennies representing the arrears of the ransom of David, king of Scotland, from York to London at an average speed of fifty-five kilometers (thirty-eight miles) a day. Riding home empty-handed, William made better than sixty kilometers (forty-five miles) a day.[156]

William was traveling alone; noblemen, kings, and prelates, who might be expected to travel fastest, were handicapped by their retinues, for whom accommodations had to be found, and rarely did better than about thirty miles a day.[157] More important for the Commercial Revolution was the improved speed of messengers carrying commodity and price information from the Champagne Fairs and other markets to home offices in Italy and, in the opposite direction, instructions to local agents.[158]

The slowest component of traffic was droves of animals. In May 1323 John the Barber set out from Long Sutton, near King's Lynn, with 19 cows and a bull, 313 ewes, 192 hogs, 172 lambs, and a bellwether, on the 130-mile journey to Tadcaster, in Yorkshire, evidently to stock the royal manors there. He had to hire a shepherd and eight boys to assist him, as well as twelve local boys "to chase the said animals through the town of Boston," a cavalcade that must have disturbed the townspeople no matter how well controlled. John covered twelve miles in the first two days, and a second twelve in one day, evidently more level going; the next eighteen miles took two days, and at the end of a week he had traveled fifty-six miles. At this point he picked up several hundred additional animals plus more boys and another shepherd. Six more days over a more direct road often used by drovers took him the remaining seventy-four miles, making his average for the whole arduous trip ten miles a day.[159]

* * *

In inland water transportation, conflicts between waterpower and navigation rights multiplied with the increasing traffic. One technological solution was the navigation weir, a small dam only partially blocking the waterway while maintaining stream depth. Its drawback, the obstacle to upstream traffic created by the strong constricted current, was dealt with by installing an animal-powered windlass to haul vessels past the dam. The Low Countries, where almost 85 percent of traffic moved on inland waterways, pioneered canal locks at Damme and elsewhere in the late fourteenth century, but early lock gates—double doors or vertical portcullis—had problems that awaited solution in the following century.[160]

Navigation: The Compass Matures

While better roads, bridges, and vehicles gradually speeded land transportation, Mediterranean shipping underwent a revolution in both technology and function beginning in the thirteenth century, doubling the number of voyages per year to Egypt and the Levant, and assuming the main burden of the trade between Italy and Flanders, heretofore carried overland via the Champagne Fairs.

Two important new types of ship contributed. The "Great Galley" introduced by the Venetian Arsenal was not a galley at all but a sailing ship that used oars for entering and leaving port.[161] Its two, later three, masts were lateen rigged, with the large mainsail supplying most of the wind power. Its hold accommodated 150 tons, silks and spices on the northern trip, wool cloth or raw wool on the return.[162]

An even more useful north-to-south carrier was a new model of the northern cog, introduced into the Mediterranean about 1300.[163] Earlier square-sailed northern ships had encountered a peculiar difficulty in the Mediterranean. The westerly wind that prevailed in the Strait of Gibraltar carried them in easily enough but virtually blocked their passage out again. Better rigging overcame that problem, and in the fourteenth century the sailing ability of the cog was given a basic improvement with the addition of a second (mizzen) mast equipped with a lateen sail.[164]

The new cog found special favor with the Genoese, who used it to carry alum, a color fixative, from the islands of Phocaea and Chios in the Aegean Sea direct to Flanders and England. While adhering to traditional carvel-skeletal construction, Genoese shipyards progressively increased the size of their hulls, by 1400 reaching cargo capacities of 600 tons, three times the size of the Hanseatic bulk carriers.[165]

The "castles," fore- and stern-, that the northern cog had added in the eleventh century were gradually absorbed in Mediterranean shipbuilding into the lines of the hull and proved as effective against the pirates of the Mediterranean as against those of the Baltic. Used as a shelter for crew and for spare rigging, the forecastle became a permanent feature of sailing ships.

The old-fashioned galley and lateen-rigged sailing ship were not completely eclipsed, despite their higher costs owing to larger crews and smaller cargo capacity. The lateen sailer was especially valuable for cabotage (coastal tramping), where maneuverability was at a premium, while the galley continued to be favored by pilgrims, for whom, in the post-Crusading world, Venice was the leading port of embarkation. The galleys' operating procedure of putting into port every night suited this class of medieval travelers, who in addition to improving their spiritual condition liked to make the most of the trip, dining and sleeping onshore and seeing the sights. On the round ships' express voyage to Syria, passengers had to carry food for the whole trip and glimpsed famous cities only from afar. Fares were lower, but conditions were steerage as opposed to first class.[166]

The maturing of the compass as a navigation instrument took place in the Mediterranean, partly because this narrow but deep sea did not permit navigation by sounding and partly because its seafarers were the most sophisticated navigators and thus were able to supply important complementary devices. The first of these was the compass card, contributed by the sailors of Amalfi and based on the ancient *Rosa Ventorum* or "Rose of the Winds." A circular card furnished with the thirty-

Model of fourteenth-century Mediterranean sailing ship has new-style castles but old-fashioned steering oar. [Science Museum, London.]

two points of the compass and positioned directly beneath the free-swinging magnetized needle fixed on a dry pivot, it allowed the helmsman to read the ship's course—in points, not degrees, since the thirty-two-point scale was incompatible with the 360 degrees of the astronomer's circle.

The second auxiliary device was the "portolan" (port-finding) chart, the world's first navigational chart. Experienced

Italian sailors felt their way on repeat voyages by sailing from one island or headland to the next, setting their course by compass and estimating the distance traveled on each bearing. A natural advance was to compile sailing directions that described coastlines and specified bearings and distances between points so that skippers unfamiliar with a given shipping route could benefit. In the late thirteenth or early fourteenth century, someone had an insight: such information could be represented geometrically with two large circles superimposed on the whole Mediterranean, one with a center just west of Sardinia, the other with a center on the Ionian coast north of Rhodes.[167]

Besides compass and charts, Mediterranean ships took to carrying hourglasses to aid in calculating ship speed and distance traveled. The astrolabe also made its appearance on board ship, again in the vanguard Mediterranean, where its value—determining latitude—was marginal.

Whether the new ship types or the new navigating techniques had more to do with the revolution in Mediterranean shipping operations that followed is a matter of scholarly controversy.[168] Both contributed to a startling change: after millennia of sailing back and forth once a year, Italy to Egypt or Asia Minor, Italian fleets took to making two such voyages. Venetian ships departed in February and returned in May, left again about the first of August and returned before Christmas. The Genoese also ceased wintering in the East and came home in time to launch a second voyage. After 1280 Pisan records too show ships sailing in all seasons, including the dead of winter.[169]

With the difficulties of winter navigation overcome, its advantage became apparent: better prevailing winds. Ships departing from Egypt in the months from May to October faced almost steady northerly and northwesterly winds, forcing them to detour around Cyprus or Rhodes, whereas in late fall the wind shifted to easterly, favoring the return to Italy. The new large ships also encouraged the Italian venture into the North Sea. The first recorded commercial penetration there was made by Genoese galleys in 1277–78; sailing ships quickly followed, and by 1314 Venetian voyages to Flanders were safe and regular.[170]

In northern waters, the value of the compass was reduced by the shallowness of the Baltic and North Seas, which permitted navigation by lead and line. But for voyages into the broad and deep Atlantic, it was invaluable, and such voyages were becoming more common for adventurous fishermen. About 1330 William Beukelszoon pioneered the practice of gutting herring at sea, improving preservation and making possible much longer fishing expeditions.[171]

By the mid–fourteenth century, the new navigation and the new ship rigging were in general use in the northern and southern seas and in the Atlantic. In 1354 Pedro IV of Aragon ordered all his ships to carry charts. By this time too, trigonometry, developed in the universities, was being applied to navigation.[172] The possibility now arose of global voyages, the unlimited exploration of all the fabled seven seas.

"The Investigation of Causes": The Scientific Attitude

Around the year 1180, a Pisan merchant was appointed to the post of customs official, or consul, of the Pisan community in Bougia, Muslim North Africa. After settling there, he sent for his son Leonardo Fibonacci,* who was still "in his boyhood" (pueritas), to complete his education "with a view to future usefulness," a commentary on the new attitude toward Islam developing among the European business class. In his new home, Leonardo made the discovery of Hindu-Arabic numerals.

Adelard of Bath's translation of al-Khwarizmi had expounded the Hindu notation but only to a very limited circle even among the mathematically literate. Leonardo perceived its enor-

---

* Leonardo's name appears in the incipits (opening lines) of his books as Leonardus Pisanus (Leonardo Pisano, or Leonard the Pisan) and as Leonardus filius Bonacii, literally "son of Bonaccio," but probably the Latinization of the surname Fibonacci. Leonardo's father is mentioned in a contemporary document as Guglielmus (Guglielmo, or William).

mous potential value and in 1202 undertook its wider diffusion by writing what proved to be a seminal book in the history of mathematics and science, the *Liber abaci* (Book of the abacus). The book began: "The nine Indian figures are 9 8 7 6 5 4 3 2 1. With these nine figures and the sign 0, any number may be written, as is demonstrated below."[173]

Many of the problems presented by Leonardo in the *Liber abaci* dealt with practical business matters, such as calculation of interest, margins of profit, percentages of alloys in coinage, and prices; others were recreational. Methods of solution were borrowed from the Hindus and the Arabs, with some refinements of Leonardo's own. His main contribution to mathematics, beyond the introduction of the Hindu numerals, was in number theory. He is recognized today chiefly as the originator of the "Fibonacci sequence," the first recursive number sequence (sequence in which the relation between two or more successive terms can be expressed by a formula) known in Europe.[174]

Leonardo's greatest achievement in number theory, however, was in Diophantine algebra, a discipline named for a fourth-century Alexandrian mathematician. Leonardo's algebra, like that of the Hindus, was rhetorical—expressed in words rather than symbols, with *res* (thing) for an unknown, *quadratus numerus* (square number) for x^2, and *cubus numerus* (cube number) for x^3. His problems, however, were accompanied by diagrams with letter labels representing the unknowns, usually *a* (alpha), *b* (beta), and *g* (gamma), prefiguring the modern x, y, and z. Signs to indicate operations did not appear until centuries after Leonardo's death—the plus and minus signs in the fifteenth century, equals in the sixteenth, division in the seventeenth.[175]

For a time businessmen were wary of the new numerals, partly out of general conservatism, partly because it was felt that they could be more easily altered by the unscrupulous, and finally because they necessitated memorizing tables of multiplication and division. But by the late fourteenth century, Hindu numerals were displacing both Roman numerals and the calculating board in European commerce. They also found their way into the literature of everyday life, although Roman numerals lingered in many places. In advanced Italy, the Datini correspondence employs the Hindu numerals and only occasionally lapses into Roman, but in more backward England a century later the letters of the Paston family still use Roman, even in dates: "Wretyn ... on the Frydaye next Seynt Symonds and Jude, anno E. iiii xix" (Written ... on the Friday after St. Simon and Jude's Day in the 19th year of Edward IV).[176] Eventually the Roman figures were relegated to secondary status, in uses such as outlines and cornerstone inscriptions.

Most significant was the impact of the Hindu notation on science and mathematics. Charles Singer calls it "a major factor in the rise of science" in the Western world.[177] The beginnings of Western trigonometry trace to the imposition of the methods of Euclid by Richard of Wallingford (c. 1292–1335) of Merton College, Oxford, on the "Toledan Tables" of the Arabic mathematician al-Zarqali.[178] Mathematics was essential to the pursuit of the study of optics, one of the favored sciences of the universities, whose clerical intellectuals were inspired or justified by Biblical citations. "In God's Scriptures," wrote Roger Bacon, "nothing is so much enlarged upon as those things that pertain to the eye and vision."[179] The Oxford master most noted for his interest in "the metaphysics of light" was Roger Bacon's mentor and one of the outstanding intellectuals of the thirteenth century, Robert Grosseteste (c. 1175–1253), in his later years bishop of Lincoln. Grosseteste perceived light as the cause of motion and the principle of intelligibility in the universe and strove to answer questions such as how the sun produces heat and how the moon influences the tides.[180] On the practical level, the invention of eyeglasses occurred in Italy sometime before 1292, facilitated by the glassmakers' mastery of the art of making clear glass. The first glasses had convex lenses, improving vision for the farsighted. Concave lenses, for the nearsighted, did not arrive until the sixteenth century.

In the universities some of the new eyeglasses were focused

Hugh of St. Cher wearing pince-nez spectacles. Detail of fresco by Tommaso da Modena, Chapter House of St. Niccola, Treviso. [Alinari.]

on rediscovered Aristotle, of whose works two thousand manuscript copies survive from the thirteenth and fourteenth centuries.[181] The other Greek "authorities" were likewise copied and recopied. The attraction of Greek knowledge lay in both quantity and form. "Arranged in neat compartments, it was presented in elegant, rational, and sophisticated fashion, and it contained an enormous amount of factual information about the natural world as well as highly developed methods of investigating that world" (Richard Dales).[182] Investigating the world was a project with immense appeal. Much as they loved Aristotle, the university scholars did not hesitate to criticize him on the basis of what they learned from their own experi-

Albertus Magnus. Detail from fresco by Tommaso da Modena, Chapter House of St. Niccola, Treviso. [Alinari.]

ence. "Natural science," said Albertus Magnus (c. 1200–1280), one of the luminaries of the University of Paris, "is not simply receiving what one is told, but the investigation of causes of natural phenomena."[183] In line with this attitude was the thirteenth-century introduction of dissection at Salerno, Bologna, and other medical schools.

In 1277 Albertus, his colleague Thomas Aquinas, and the rest of the Paris faculty received a shock in the condemnation by the bishop of Paris of a number of their teachings, including the daring one that God had no power to move the world with a rectilinear motion, the universe's motion being invariably curved. But the masters soon found a way around the objection. God could do anything, granted, but under the agreement of "ordained power," he produced the world and cosmos as they actually exist, in conditions that preclude straight-line motion.[184]

Despite this contretemps, the Church's attitude toward scientific inquiry remained benign. Its opposition to the two great "false sciences" of alchemy and astrology was mild. In alchemy, it condemned the charlatanism often practiced on the gullible, such as that perpetrated on the priest in Chaucer's "Canon's Yeoman's Tale," victimized by a trickster who pretends to turn base metal into silver. The Church approved the study of "the transmutation of the metals, that is to say, the imperfect ones, in a true manner and not fraudulently."[185]

In astrology—"applied astronomy," in the apt phrase of Forbes and Dijksterhuis—the Church did not officially tolerate the casting of horoscopes, which seemed to conflict with the doctrine of free will, but the Paris intellectuals lent a cover of sanction to the popular pseudoscience by arguing (exactly as had Ptolemy a thousand years earlier) that the individual could evade his star-predicted fate by the proper conduct of his life.[186] At Oxford, Robert Grosseteste first accepted planetary influences on human life—if the moon could affect the tides, why could not the planets influence human beings?—but later rejected it, first on the ground that current astronomical instruments were not precise enough to permit reliable judgments, and second because "the free choice of a rational mind is subject to nothing in nature, save only God—in short . . . it was bad science, and it was bad morality" (John F. Benton).[187] Nevertheless, horoscopes continued to be cast based on the positions of the heavenly bodies at the hour of the subject's birth.

As from early times, astrology played a role in medicine. The ancient Chinese emperors' "pernoctation rota"—the sequence in which each month they slept with their different categories of spouses and concubines—had demanded accurate astronomical observations, in the interest of the imperial succession; similarly, in 1235 Holy Roman Emperor Frederick II consulted his astrologer as to the optimum time for consummating his marriage to Isabella, sister of English king Henry III. Besides aiding rulers to produce suitable heirs, astrology had an everyday function for the physician. Chaucer describes his "Doctor of Physic":

> No one alive could talk as well as he did
> On points of medicine and of surgery,
> For being grounded in astronomy,
> He watched his patient's favorable star
> And by his Natural Magic knew what are
> The lucky hours and planetary degrees
> For making charms and magic effigies.[188]

Astrological influences were sought to explain the Black Death and attempts made to predict future plagues on the basis of planetary conjunctions, credited with causing corruption of the air.[189]

The instruments of astrology, however, were used effectively for genuine astronomy. With the astrolabe, armillary sphere, and a few other simple tools, Guillaume de St. Cloud, a follower of Roger Bacon, founded the school of astronomy of the University of Paris, where one of his first accomplishments was the correct determination of the latitude of Paris (48° 50'), a demonstration that showed the potential of such instruments in navigation.[190]

Unlike his Greek and Roman forebears, the thirteenth-century intellectual included the "banausic arts" in his wide-ranging field of interest, as indicated by Roger Bacon's description of his friend Pierre de Maricourt: "He knows by experience the laws of Nature, Medicine, and Alchemy . . . He has delved into the trade of the metal founders. He has learned everything concerning warfare, weaponry, and hunting [as well as] agriculture, surveying, and the work of the peasants . . . [and also] the procedures of the old witches, their spells . . . everything concerning magic, and also the tricks of the jugglers."[191] Maricourt (also known as Petrus Peregrinus—Peter the Pilgrim) was one of the most tireless pursuers of perpetual motion, a chimera imported from India that drew the efforts of many, including Villard de Honnecourt, who sketched a perpetual-motion device in his notebook. The still-unlocked mysteries of wind, tides, and rivers lent a plausibility to the will-o'-the-wisp that was pursued long after the Middle Ages. More usefully, Pierre de Maricourt

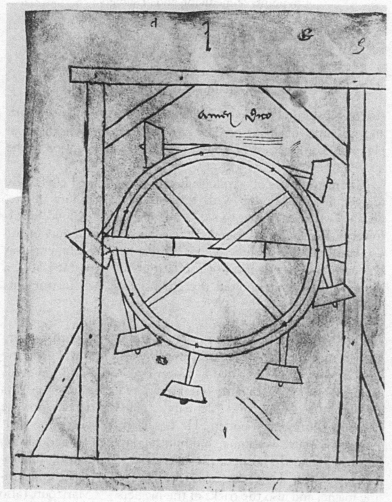

Perpetual-motion machine sketched by Villard de Honnecourt. "Often have experts striven to make a wheel turn of its own accord," reads the caption. "Here is a way to do it with an uneven number of mallets and with quicksilver." [From The Sketchbook of Villard de Honnecourt, ed. by Theodore Bowie, Indiana University Press.]

studied magnetism intensively and wrote an important treatise on it.

Roger Bacon's own principal work was an ardent advocacy of the reform of the educational system to emphasize experimental science and mathematics. Even St. Francis of Assisi, the illustrious founder of the Franciscan Order, to which Bacon and Grosseteste belonged, was a contributor to the new scientific spirit. "It may be said . . . that St. Francis first taught Europe that nature is interesting and important in and of itself" (Lynn White).[192]

By the thirteenth century, speculative thought was no longer confined to the clerical intellectuals. Outstanding among a handful of prominent lay figures were two sovereigns, Alfonso the Learned of Castile (ruled 1252–1284) and Frederick II (ruled 1220–1250). Alfonso caused scientific complications to be drawn up and foundations laid for almanac and calendric calculations. Frederick's scientific curiosity about the natural world and pursuit of its mysteries won him the nickname "Stupor Mundi," or Wonder of the World. He studied chicken embryos, maintained a menagerie of exotic animals, caused Arabic scientific works to be translated, corresponded with Muslim potentates about mathematics and philosophy, and wrote a book on falconry, noting that "Aristotle has rarely or ever had experience in falconry, which we have loved and practiced all our lives."[193] In the 1220s he paid a visit to Leonardo Fibonacci in Pisa, setting the mathematician difficult problems whose solutions Leonardo included in two of his subsequent books. A prominent layman of the fourteenth century with a scientific bent was Geoffrey Chaucer, who in 1391 wrote a treatise, *The Astrolabe*, for his young son bound for Oxford.[194]

At the end of the thirteenth century, its most extraordinary literary and scientific project appeared, Marco Polo's *Description of the World*, depicting to a wondering Europe the vast size and wealth of China. Unfortunately both for contemporary Europe and for scholars today, Marco skimped on descriptions of Chinese technological accomplishments, omitting mention, among other things, of the Great Wall. But the enormously popular book (over eighty fourteenth- and fifteenth-century manuscript copies, in several languages, survive) contributed to the growing dream of reaching Asia by a direct sea route. An attempt had already been made. In 1292 a pair of Genoese galleys rounded the Strait of Gibraltar southward and were never heard from again. Other Genoese mariners, in the service of Portugal and Spain, kept the idea alive; an interim reward was gathered in 1336 when the Canary Islands, the "Fortunate Isles" of the ancients, were rediscovered by a Portuguese expedition commanded by the Genoese captain Lancelotto Malocello.

But the location and shape of the southern tip of Africa remained speculative, and an alternative strategy for reaching the East—sailing west—was equally so. Roger Bacon, Albertus Magnus, and their colleagues brought a renewed emphasis to the sphericity of the earth, never questioned but somewhat lost sight of in the Crusading age, which had popularized a kind of stylized map showing Jerusalem as the center of the world. Roger Bacon joined Marco Polo in exaggerating the breadth of Asia and underestimating the westward distance from Europe to China. His views were cited by Pierre d'Ailly (1350–1420), whose *Imago mundi* (Image of the world) was a principal source for Columbus's calculations.[195]

With the earth's sphericity taken for granted, fourteenth-century scholars began focusing on the question of motion. Jean Buridan (1300–1358) of the University of Paris showed that appearance of motion is relative and that whichever moved, the earth or the universe around it, appearances would be the same. Two ships, alone within sight of each other, with no other visible point of reference, could not tell which was moving. Buridan's colleague Nicolas Oresme (c. 1325–1382) argued for the rotation of the earth on its axis to explain the diurnal motion of the heavens and resolved a paradox that argued against the theory: an arrow shot straight up in the air comes down in the same spot despite the earth's movement beneath it. The arrow, Oresme explained, was moving laterally at the same speed as the earth before it left the bow and continued to do so while appending its perpendicular flight.

Once intellectuals began asking such questions and seeking answers, they were embarked on a path of speculation and experimentation down which lay much sharper collisions of science and faith, and also vistas of knowledge that the fourteenth century could hardly envision.

The tonsured, Latin-speaking clerical intellectuals of the Middle Ages bequeathed to the universities of the future the function of scientific research. Yet teaching was always their priority and, furthermore, teaching with a material value. Charles Haskins quotes a letter from a father to a son that sounds the note of a later day. Remarking that his proposed course of study for the priesthood would cost "a great deal of money," the father tells the son that he would be "better advised" to take up physics or medicine or "another lucrative science."[196]

EUROPE 1400

The fourteenth century ended on a somber note; the Black Death paying one of its return visits to a Europe that had not yet recovered from its earlier devastations. Population, urban and rural, was still below its preplague level, the Hundred Years War smoldered in a precarious truce, and the Church was riven by the Great Schism, which enthroned rival popes in Rome and Avignon. Yet under the surface, and despite calamities both substantive and superficial, Europe had advanced to a point where it at last rivaled Asia as a center of civilization. In power sources, industrial organization, architecture, shipbuilding, and weaponry, it had absorbed its many borrowings and synthesized them with its own inventions to create a technical apparatus far beyond that of the ancient civilizations that gave it birth.

For better and for worse, modern nations and modern society were taking shape. European workers were laboring by the clock; European intellectuals were prying into the secrets of the universe. Technology had won a position of esteem altogether new. Personified by the master mason in the construction yard of the rising cathedral, the skills that Aristotle had disdained as too commonplace to be worthy of study were turning out to possess mysteries and promises to intrigue the most inquiring intellects.

7

LEONARDO AND COLUMBUS

THE END OF THE MIDDLE AGES

ROBERT S. LOPEZ SPEAKS OF "THE TWILIGHT OF the fifth century," "the dawn of the tenth," and "the glare of the fifteenth."[1] Where scholars grope for information in the scanty records of the early period, in the fifteenth century they are engulfed in a flood of documentary sources, including, from the 1450s on, printed materials.

Once upon a time, a historical theory was formulated and taught to generations of students to explain the transition from medieval to modern times: a crowd of intellectual refugees from the fall of Constantinople in 1453 brought Greek learning to the West and launched the "Renaissance." The theory, never very convincing to medievalists, then few in number, was exploded beyond recovery by Charles Homer Haskins's *The Renaissance of the Twelfth Century*, published in 1927, establishing that Greek science came to the West largely through the agency of the Arabs and that the accomplishments of the European fifteenth century were the continuation of a long process.[2]

Nevertheless, the fifteenth century has an unmistakably revolutionary demeanor, with its gunpowder artillery, its printed books, its transoceanic voyages, and its flowering of art. The many-sided development at long last carried Europe past Asia to world leadership in technology, at a moment when one element of that advance, the full-rigged ship, was suddenly bringing Europe, Africa, Asia, and the brave New World together in a cultural collision unique to human history.

The most impressive documentation of Europe's rising creative powers may be found in the notebooks of Leonardo da Vinci (1452–1519). During their successive rediscoveries in modern times, these astonishing collections were mistakenly perceived as sketches of original inventions, products of an individual "Renaissance" genius. The misconception was due in part to the aesthetic quality of the drawings, and in part to a prevailing notion of the character of invention, an exaggeration of the contribution of the individual "inventor" and an underappreciation of the social nature of technical innovation. The historic value of Leonardo's "notebooks"—actually an immense scattering of sketches and jottings—lies less in their author's own contributions to engineering than in their incomparable illustration of the atmosphere in which he lived, a time in which dreamers, tinkerers, and artist-inventors were applying themselves on the frontiers of technology opened by the discoveries of their medieval predecessors. In Bert S. Hall's words, the sketches "tell us about the processes of invention and the manner in which available techniques of the later Middle Ages and Renaissance could be put together in novel patterns."[3]

Again and again, Leonardo's conceptions echo those not only of his peers but of his predecessors, sometimes of a much earlier era. His famous sketch of an ornithopter, or flying machine, was preceded not only by earlier speculations about flight but by an actual attempt, made in the eleventh century by an English inventor named Eilmer, recorded by chronicler William of Malmesbury:

> He was a man learned for those times, of ripe old age, and in his early youth had hazarded a deed of remarkable boldness. He had by some means, I scarcely know what, fastened wings to his

hands and feet so that, mistaking fable for truth, he might fly like Daedalus, and, collecting the breeze on the summit of a tower, he flew for more than the distance of a furlong. But, agitated by the violence of the wind and the swirling of air, as well as by awareness of his rashness, he fell, broke his legs, and was lame ever after. He himself used to say that the cause of his failure was his forgetting to put a tail on the back part.[4]

Other visionaries who drew or described flying machines all slipped, like Leonardo and Eilmer, into the ornithopter error, the imitation of wing-flapping birds. Leonardo's pyramidal-shaped parachute was also anticipated, though only on paper (and more recently) by sketches dated between 1451 and 1483. Leonardo's chute looks more practical, but nobody ventured to try out the idea for another three hundred years.

Not only the inventions but the art and literature that are the most famous products of the Renaissance were firmly rooted in the Middle Ages. Giotto, Dante, Petrarch, Boccaccio, and Chaucer all lived before 1400, and the last four all owed something to the work of forebears, from the Provençal poets to the authors of the fabliaux.

Renaissance artists profited from the Commercial Revolution's creation of the affluence that underwrote art. When he was in Avignon, Francesco Datini treated artistic works like any other of the commodities he dealt in, ordering paintings from Florence by subject matter and size:

> [Send me] a painting of Our Lady with gold background . . . by the best master who is painting in Florence, with several figures. In the middle, Our Lord on the cross or Our Lady, whichever you find . . . with fine big figures, the fairest and best that you can find for 5½ to 6½ florins, but not more . . . A picture of Our Lady of the same kind, on gold background, but a little less grand, for the price of 4½ florins, but no more.

Later one of his partners wrote to the correspondents in Florence, "Pictures are in no great demand here; they are occasional items which one must buy when the painter is in want."

Artists, Francesco thought, had too high an opinion of themselves. "Are they all brothers or cousins of Giotto?" he inquired sarcastically.[5]

Columbus's voyage, despite its theatrical circumstances, grew out of the slow maturing of the full-rigged ship, with its complement of navigational instruments, and their inevitable application to the search for a sea route to Asia. European perception of the world outside the Mediterranean and Baltic basins was gaining focus as contacts multiplied. Delegates from the Ethiopian Christian Church attended the Council of Florence in 1441. Ethiopia, on the Gulf of Aden, was halfway to India. In 1487, the same year the Portuguese government sent Bartolomeu Dias off to find the Cape of Good Hope, and another expedition westward from the Azores to explore the Ocean Sea, it dispatched Pero da Covilhã overland via Egypt to gather information on the Indian Ocean.[6]

Besides ocean navigation, two other technical systems with roots in the earlier Middle Ages came to the fore in the fifteenth century. The first was the assembly of paper, press, ink, and type into letterpress printing, which at once made a torrent of information on a range of subjects available to a wide public, in fact, the entire literate Western world. The second was the improvement of gunpowder weaponry into the effective firearms that conferred on Europeans an advantage that has been exaggerated but was nevertheless significant in their sudden confrontation with the rest of the world.

In the political sphere, the fifteenth century witnessed the emergence of several large national states, not only in western but in northern and eastern Europe. As Poland, Russia, Sweden, and Brandenburg-Prussia squeezed the Hanseatic League out of its power and privileges, German ships ceased to dominate the northern seas, while the central European metal mines assisted the rise of the Habsburg dynasty.

The swarm of Italian city-republics, led by Venice, Milan, Florence, and Genoa, competed ferociously for political and commercial advantage, at first oblivious of the rise of a new power: the Ottoman Turks, who captured the world's attention along with Constantinople in 1453. Thenceforward the Turks contested the eastern Mediterranean with the Europeans, contributing to the motivation for a new route to Asia.

Crowded and turbulent, only in retrospect could the fifteenth century be seen for what it was, the hinge by which the medieval Western world moved forward into modern times.

"The Admirable Art of Typography"

Half a century after the event, Johann Schoeffer, son of Gutenberg's assistant Peter Schoeffer, in a preface to a newly printed edition of Livy, wrote unequivocally: "The admirable art of typography was invented by the ingenious Johann Gutenberg in 1450 at Mainz."[7] Later the younger Schoeffer reneged and claimed the invention for his father, but there is little doubt that the credit has always properly been awarded to Gutenberg. A lawsuit against the inventor by his financial backer Johann Fust, Peter Schoeffer's father-in-law, contains what most scholars feel is conclusive evidence.

There are, however, qualifications attached to Gutenberg's title. The Asian priority of invention of movable type is now firmly established, and that the Chinese-Korean technique, or a report of it, traveled westward is almost certain, though the path of transmission remains unknown. Of the four major components of letterpress printing, paper already existed, and the other three—press, type, and ink—were original with Gutenberg only in a narrow sense. Printing, in fact, has often been cited as a good illustration of the social character of invention.

In the late fourteenth century, following the introduction of xylography—wood-block printing, probably learned from China—a popular art form sprang up, a religious picture with some brief text added, carved in wood and printed. In the Low Countries and the Rhineland, booklets known as "Poor Man's Bibles" were mass-produced, along with playing cards (newly invented), posters, calendars, and short Latin grammars called "donats," from the Roman grammarian Aelius Donatus.[8]

Out of the wood-block donats and playing cards, before the invention of movable type, came that of engraving, the incising of copper plate with a chisel-like implement (burin or graver), permitting the reproduction of many more copies than were possible with a woodblock.[9] By midcentury, engraving was an established technique in south Germany and north Italy on its way to becoming an independent art form.

Meanwhile Asian printing was reaching maturity in China and Korea, where wooden type was supplanted by bronze in the early fifteenth century; shortly after, in Europe, a similar progression took place. Wood type was first employed by bookbinders to stamp titles on manuscript bindings,[10] and as early as the 1420s, Laurens Janszoon of Haarlem in the Netherlands may have experimented with it for general use.[11] It soon became clear to European printers, as it had to Asian, that wooden type was not satisfactory in either uniformity or durability. Gutenberg, by trade a silversmith, and other metalworkers realized that their die-cutting technique could be applied to block printing via a clay matrix on which the text was struck letter by letter; the whole page, in relief, was then cast in lead. Gutenberg employed the technique in Strasbourg, where he had temporarily relocated from Mainz; it was also practiced in the Netherlands and the Rhineland.

Thus experimenters were converging on the final step to movable type. Alignment gave problems, and the process of striking each letter in clay threatened to deform the neighboring letter. In 1426 the combination of die cutting, matrix, and lead casting suggested to Gutenberg the potential of individually cast metal letters.[12]

Spectroscopic analysis shows early type metal to have been an alloy of lead, tin, and antimony, a combination unchanged in centuries of letterpress typography. At first the dies were also soft metal, but in the 1470s these were replaced by steel, an improvement credited to Peter Schoeffer, Gutenberg's assistant (and Johann Schoeffer's father). Including joined letters and symbols, each typeface in upper and lowercase required about 150 characters. For each a steel punch or die was first cut and employed to form a mold in which the soft-metal type character was cast. Both punch and type had to be finished by hand filing, more difficult with the hard steel punch, which, however, lasted much longer than the soft type.[13]

Typesetting was also laborious, the typesetter picking each character from the composing stick with tweezers, setting line by line in the chase, and taking and correcting proof. Each page of the Gutenberg Bible probably took one man one day. Proofreaders scanning the Latin text naturally indicated their corrections in Latin, a custom whose vestiges remain, as in "stet" for "let it stand."[14]

Gutenberg's typeface imitated "Gothic," a thick "black-letter" script developed out of Caroline minuscule in the tenth century. In Germany the style was popular and remained in vogue until modern times, but in Italy it was perceived as out of step with the new classical spirit. In 1465 two German printers in Subiaco invented a typeface for an edition of Cicero that developed into what subsequent generations called "Roman." A few years later printers in Venice introduced "italic." The new faces were cut with a skill often amounting to art, as in the Roman face created by Nicolas Jenson, a French printer in Venice who had studied under Gutenberg.[15]

The standardization of individual characters imposed by movable type proved an added benefit: the uniformity of product often regarded as a negative feature of mass production was a positive aid to the reader of books.[16]

The commercial success of block printing had meanwhile stimulated another key development. Printing required ink with different characteristics from those of water-based writing ink, which smudged, refused to spread uniformly, and showed through on the reverse side. Italian painters in the previous century had invented the technique of mixing pigments (insoluble natural substances) suspended in linseed oil. Gutenberg experimented successfully with a mixture of lampblack (soot recovered from chimneys), turpentine, and linseed or walnut oil. Reduced by heating, the new ink shone black and adhered to slightly dampened paper without blurring.[17] The old oak-gall and iron mixture continued in use for writing, as Shakespeare notes in *Cymbeline*: "I'll drink thy words . . . though ink be made of gall."[18]

Along with type and ink, Gutenberg had another flash of insight, that the ancient wine-and-oil press, already modified for paper manufacture, required only a slight further change to assume a function in printing (Chinese printing did not employ a press, relying on the rubbing technique used for woodblocks). Once its configuration was worked out, the wooden screw press, consisting of a sliding flat bed and an upper platen, could be operated rapidly to produce a sharp impression. Its steeply pitched screw required only a quarter turn of the lever, and the sliding bed allowed easy inking. Two-color and multicolor printing were introduced in the very first printed works, the celebrated forty-two-line Bible of Gutenberg and the Psalter which

Model of Gutenberg press. [Science Museum, London.]

Fust and Schoeffer, after winning their lawsuit against the inventor, produced with his equipment. The red second color of the Bible and the multiple colors of the Psalter were achieved by the simple process of removing from the composed page characters or passages to appear in color, inking them separately, and returning them to the press, so that all colors were produced in a single impression. Problems of register (overlapping of colors) were thus avoided at the cost of some production time.[19] Engraving was quickly absorbed into the printing process, though a number of early printed works in small editions employed hand illumination, copy by copy, for their decoration.[20]

Two years after the appearance of Gutenberg's Bible in 1455, the first printing press in Italy went into operation, followed by others in Paris and London (Caxton's), and by 1480 nearly every city in Europe had at least one press.[21] The economics of the invention were irresistible. A Florentine scribe could produce a copy of Plato's Dialogues for one florin; a press charged three florins per quinterno (five sheets, or eighty pages of octavo) for typesetting and printing, and could produce an unlimited number at the cost of paper and ink.[22]

The earliest print shops were much like those still operating in Mark Twain's day: cases to hold the type; a table with shelves of blank leads and room for composed pages; composing stick and galley; a copyholder; tweezers to handle the type; the press stone on which pages were prepared for printing; the chase in which pages were tightened into rigid form with the aid of wooden wedges; an inkpot and ink pads; a table to hold blank sheets and receive the printed pages; a tub of water to dampen the sheets to improve the impression; and a stretcher to hang the freshly printed sheets to dry.[23] Some of the very earliest practitioners established the tradition of the itinerant printer, carting equipment from place to place. Many of the first printers were former priests, whose knowledge of Latin was helpful; many editors and proofreaders were former abbots.[24]

Venice emerged as the printing capital of Europe, publishing 2,789 books by 1500. Altogether, the "incunabula" ("newborn"), the name given to books published before 1500, numbered some 40,000 editions of various works totaling 15 to 20 million copies.[25] Most (77 percent) were in Latin, and almost half (45 percent) were religious works (94 Bibles in Latin, 15 in German, 11 in Italian). But the early medieval encyclopedias of Isidore of Seville, Cassiodorus, and Martianus Capella, as well as the later encyclopedia of Bartolomaeus Anglicus, were issued in printing after printing through the sixteenth century. Many agricultural treatises, mostly Roman, were printed. Boethius's work on arithmetic (De arithmetica) was printed repeatedly from 1488 on.[26] In 1493 Columbus's letter to the king and queen of Spain, detailing his discovery, was printed (translated into Latin) and by the end of the year was circulated in eleven editions to an eager European public.[27] Before or shortly after 1500, the works of Robert Grosseteste, Albertus Magnus, Thomas Aquinas, Roger Bacon, and most of the other leading medieval writers on scientific subjects reached print. A few contemporary works were printed, including two arithmetical texts that employed Leonardo Fibonacci's Hindu numerals, contributing to the development of algebraic notation and the rise of mathematics.[28]

Experience led printers to smaller typefaces and standardized book sizes, as for the first time in history a large educated class was given access to a wide array of knowledge. The literate elite, once all but limited to the educated clergy, now included not only the nobility, many of whose sons attended the universities, but a large and growing number of lay commoners, members of the middle class created by the Commercial Revolution. Frankfurt and other cities organized book fairs, and peddlers hawked the latest editions through the towns. If printing was born in Asia, the honor—and profit—of turning it into an efficient mass-production process, a democratic form of communication, belongs incontestably to Europe.

Of all the superlatives accorded to the invention through the centuries since Gutenberg, one of the most acute is that it represented "a technological advance which facilitated every technological advance that followed it" (Derry and Williams).[29]

Guns and National States

If the printed book was the most "admirable" innovation of the fifteenth century, the firearm, now reaching maturity after a slow start, was the most dramatic. Erratic black powder was tamed to consistency by the invention in the 1420s of "corning," or granulation, by which the powder, dampened by vinegar, brandy, or "the urine of a wine-drinking man," was passed through a sieve, forming coarse granules, not only safer to handle but more reliable in action. Experimentation with mixtures improved explosive power, and consequently range and accuracy. Gradually the weight of the projectile diminished in proportion to the weight of the gun, and the weight of powder rose in proportion to the projectile.[30] The hot-wire ignition was replaced by the slow match, a cord soaked in niter and alcohol, dried, and set aglow in preparation for combat. The touchhole of the hand weapon was moved to the side and a priming pan added. In about 1425 an S-shaped device called a "serpentine" was provided to hold the slow match; pressing one end brought the match into contact with the priming powder, making the gun an effective one-man weapon.

The potential of such a gun, sometimes called a "culverin," was illustrated at the siege of Orléans in 1429. French knights were attempting to storm the Augustins, an English-held fortress across the Loire, and a "big, strong and powerful Englishman," according to Joan of Arc's squire Jean d'Aulon, was valiantly resisting. D'Aulon pointed him out to a gunner named Jean de Montesclere, who aimed his weapon, fired, and caused the big Englishman to topple over, opening the way to capture of the stronghold.[31]

Late in the century, the firing mechanism was enclosed and given a spring trigger; when a wooden stock was added to absorb the recoil, the matchlock musket was complete. Its successors, the wheel lock and the flintlock, appear in some fifteenth-century sketches, including those of Leonardo da Vinci, but were not employed on the battlefield for another century.[32] The new weapon had its shortcomings compared with the bow or crossbow, including mechanical failure, problems with wet weather, and reloading, which took several minutes during which the musketeers had to be protected by pikemen interspersed among their formations. Nevertheless, by 1500 the musket, or arquebus, was making its presence felt on the battlefield, and over the next few decades it succeeded in supplanting the powerful steel crossbow.

Similarly, gunpowder artillery crossed a threshold. Fabrication was made easier by a new technique, casting in a mold to form a hollow cylinder around a mandrel (core); using the same mold guaranteed identical calibers.[33] In the closing stage of the Hundred Years War, the royal French artillery under the command of the Bureau brothers, a pair of talented smiths, used iron cannonballs to batter down one English-held castle and town wall after another and even performed effectively in the field, as at Castillon, the war's last battle, in 1453.

That same year Ottoman sultan Mahomet II bombarded

A fifteenth-century siege: the cannon at lower right lacks a gun carriage, and fires stone cannonballs. [Wavrin, Chroniques d'Angleterre, British Library, Ms. Royal 14 E IV, f. 59v.]

Constantinople with giant cannon fabricated at Adrianople by Orban, a Byzantine defector. The largest, called "Mahometta," hurled stone cannonballs of 1,000 pounds and needed more than 100 soldiers to maneuver it. Its detonation was said to cause miscarriages among pregnant women, an inconvenience that proved short-lived; "Mahometta" cracked on its second day of action and had to be abandoned.[34] Orban's other cannon were more effective. Loaded on wagons and hauled to Constantinople by teams of thirty pairs of oxen, the fifteen-ton bronze guns were rested on the ground and chocked up by stones for firing elevation.[35]

The Turks' old-fashioned stone-shooting big cannon were the wave of the past. That of the future lay in the smaller, more maneuverable, more numerous iron-shooting European artillery.[36] The Bureau brothers improvised a primitive version of the gun carriage, which was improved in the second half of the century with spoked and dished wheels, and finally by the introduction of trunnions, forming a cradle that permitted the muzzle of the gun to be raised or lowered and also absorbed some of the recoil.[37] Thenceforward cannon were a regular feature of battle as well as siege. The famous condottiere Bartolommeo Colleoni (1400–1475) introduced a new tactic, training his infantry to open gaps through which the artillery, stationed behind it, could fire. In 1475, at the siege of Burgos, in Spain, the last employment of the old catapult artillery was recorded.[38]

Gunpowder weapons had three conspicuous effects. First, artillery reinforced the trend toward the national professional army, since only a wealthy central government could afford it. Sovereigns often took a personal interest in their martial toys; John II of Portugal and Emperor Maximilian of Germany were two who expressed not only enthusiasm for but genuine expertise in the "art of gunnery."[39] Second, small arms made the armored knight obsolete, not so much because his armor did not stop musket balls as because the new musket infantry was cheaper to arm and equip and more flexible to employ, and the emerging pistol-armed cavalry of the sixteenth century much more formidable. Breastplates and helmets continued in fashion through the seventeenth century, but chain mail and full armor disappeared except for parades and tournaments. Individual prowess, hallmark of the age of chivalry, was curtailed as the new-model army extended the principle of standardization from arms and ammunition to uniforms and drill. Third, the curtain-walled castle was superseded by the low-profile, thick-rampart fortress, capable of absorbing the shock of heavy cannonballs and furnishing a good platform for defensive artillery but ill adapted to service as a private residence.[40] The new-style fortifications mostly supplanted old-fashioned city walls and were manned by garrisons belonging to the central government. The aging castles of the feudal nobility sank to the status of not very comfortable country houses, storage depots for gunpowder and cannonballs, or prisons for distinguished captives.

Joseph Needham points to China's influence in the large social changes at both ends of the European Middle Ages: "Thus one can conclude that just as Chinese gunpowder helped to shatter this form of society at the end of the period, so Chinese stirrups had originally helped to set it up." Neither invention had any perceptible impact on Chinese society, owing, in Needham's interpretation, to its relative stability compared with Western society.[41] However that may be, the origins of feudalism in Europe involved much more than stirrup, horseshoe, and saddle, and, by the same token, feudalism was already in decline when gunpowder gave it a final push toward the grave by benefiting national governments at the expense of the old castle-building, armor-wearing, horseback-riding feudal aristocracy.

A subtler effect of the new weaponry and fortifications was their impact on the incipient engineering profession. Expertise was suddenly in great demand. In response, technical treatises began to appear. The first important one came from southern Germany, where metal mining contributed to the growth of an arms industry. The *Bellifortis* (Strong war) of Konrad Kyeser of Eichstadt (1366–after 1405) remained a bible for military leaders for more than a century. Kyeser has been called "the first great engineer who has left us a well-established technological oeuvre" (Bertrand Gille).[42] A physician by profession, Kyeser published his work at the beginning of the fifteenth century, when the gunpowder age was still new. Among his sketches are a battery of cannon mounted on a turntable to be fired in succession, an artillery-carrying chariot, and a long-barreled, small-bore culverin resting on a stand. But of an array of proposed war chariots armed with pikes, lances, scythes, and hooks, only two carry rudimentary cannon, and the incendiary projectiles Kyeser sketched were ammunition not for guns but for crossbows.

At about the same time as *Bellifortis*, another military treatise, of a different coloration, appeared. Cleric Honoré Bonet's (or Bouvet's) *The Tree of Battles* set down on paper the unwritten "law of arms" accumulated by the knightly class over the previous centuries. Accepting war as inevitable, Bonet attributed its evils and injustices to "false usage, as when a man-at-arms takes a woman and does her shame and injury, or sets fire to a church." Civilian populations should be respected,

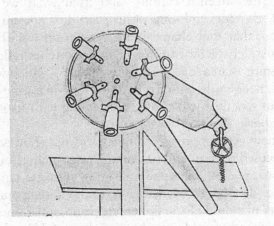

Sketch of battery of cannon mounted on a turntable, to be fired in succession, from Konrad Kyeser's Bellifortis (c. 1405). [Tiroler Landesmuseum Ferdinandeum, Innsbruck.]

for "the business of cultivating grain confers privileges on those who do it . . . In all wars poor laborers should be left secure and in peace, for in these days all wars are directed against the poor laboring people and against their goods and chattels. I do not call that war, but pillage and robbery." Bonet sought to combine the chivalric tradition with the new military age, admonishing his reader to defend "justice, the widow, the orphan, and the poor," while accepting discipline, obeying orders, and avoiding impulsive, individualistic action. The modern soldier owed his loyalty "first to the king, then to his lord, and finally to the captain." He must always remember that his actions were performed "as a deputy of the king or of the lord in whose pay he is."[43]

Knights-errant had ridden into the sunset. In their place were professional soldiers, who "followed their mercenary calling / And took their wages and are dead."[44]

The Artist-Engineers: Leonardo da Vinci and Company

Like *Bellifortis*, *The Tree of Battles* enjoyed great popularity through the fifteenth century, and from what we know of Leonardo da Vinci and some of his peers, it is not unreasonable to suppose that they shared Bonet's views on war. The rather bloodthirsty character of some of their sketches reflects less a martial spirit than a love of gadgetry, much of which was, at the moment, necessarily military. Europe's long-maturing mechanical genius was neatly, if fatally, converging with its graduation from medieval to modern warfare.

The new generation of engineers was not, however, exclusively focused on war. Even Konrad Kyeser included in *Bellifortis* siphons, pumps, waterpowered mills, furnaces, and baths.[45] Leonardo da Vinci's predecessors and contemporaries were inheritors not only of Guido da Vigevano and Konrad Kyeser but of two other, older traditions, that of Villard de Honnecourt and the master masons of Gothic architecture, and that of Roger Bacon, Jean Buridan, and the other clerical intellectuals fascinated by the secrets of the natural world.

The most striking aspect of the work of the fifteenth-century artist-engineers, abundantly illustrated in Leonardo's notebooks, is the rich redundancy in which ideas for mechanical invention now multiplied, new ideas for performing old functions, ideas for new functions, alternate approaches to problems, new applications of known principles, new combinations of familiar components. Noteworthy too is a spirit that emerges—a spirit of enjoyment, of amusement even, reminiscent of the toy mechanisms with which Heron of Alexandria entertained his circle. The wing-flapping aircraft, the helicopterlike whirlybirds, the parachutes breathe the spirit of Johan Huizinga's *homo ludens*, man at play.[46] It was a spirit that reached beyond the engineers. The mechanically minded aristocracy of the seventeenth and eighteenth centuries was foreshadowed in the fifteenth; Emperor Maximilian was one of several dignitaries who employed the wood lathe as a toy.[47]

Yet if the artist-engineers were dreamers, they were also serious thinkers, concerned, beyond the practical applications of their devices, with the large questions that had long occupied the university scholars. Leonardo's restless curiosity drove him from studying anatomy by dissection to interviewing workmen on the Visconti canal in Milan. He echoed Albertus Magnus in his insistence on firsthand knowledge: "To me it seems that all sciences are vain and full of errors that are not born of experience, mother of all certainty."[48]

Villard de Honnecourt had pioneered the combination of art and engineering, and of the transmission of technology by document in place of the age-old oral-and-manual tradition. The fifteenth-century artist-engineers carried out the revolution Villard had signaled, coincidentally just as printing arrived on the scene. Thus technology passed almost overnight through two shifts of medium, first from oral to written and drawn, and second from manuscript to print. An important social result followed: "The illustrated treatise and its printed descendants," writes Bert S. Hall, "fostered contacts between the technician's work and the world of high culture" as the treatises found an enthusiastic audience among the educated elite.[49]

The authors of the treatises were themselves neither aristocrats nor peasants but members of the rising middle class. Leonardo sprang from a long line of notaries and was apprenticed to a goldsmith; Leon Battista Alberti was the son of a new-rich banking family; Paolo Toscanelli of a family of silk-and-spice merchants.[50] Bourgeois intellectuals par excellence, they differed in their artistic abilities but shared the improvements in expression recently achieved by painters: Flemish realism and Italian linear perspective, far more effective for their purposes than the old medieval illuminators' style. Leonardo sometimes sketched a sequence of variations on the same device, suggesting that he improvised as he drew, something others may have done as well; thus the drafting pen became a tool for inventing, perhaps aided by the availability of cheap paper.[51]

Among the best known of Leonardo's predecessors are these:

Filippo Brunelleschi (1377–1446), best remembered as the architect of the brick dome of the Duomo of Florence, one of the great Renaissance architects, and inventor of many mechanical devices. He took a creative interest in problems of statics and hydraulics, in mathematics, and in clockwork. He also pioneered patent protection for inventors, receiving from the Republic of Venice the first patent ever awarded.[52]

Mariano di Jacopo Taccola (1382–before 1458), usually known as Il Taccola, one of a number of outstanding engineers of Siena, a small but combative rival of Florence. Called "the Sienese Archimedes," Il Taccola was known primarily as a military engineer but was a thinker of wide interests as well as a talented painter. Informed rather than innovative, he had a command of existing technology from which Leonardo borrowed. His two books, *De ingeneis* (On inventions) and *De machinis* (On machines), contain many hydraulic devices, including a mill driven by water falling from a tank kept full by pumping: the miller could pump in the morning and do other chores for the rest of the day while the mill operated, a sort of early version of today's pumped-storage power technology.[53] Taccola also sketched a caisson, a device revived from Roman bridge engineering: a double-walled box lined with concrete sunk in the stream and filled with rubble as part of a pier foundation.[54]

The author of the "Hussite War Ms.," an unusual anonymity among the artist-engineers, identified only as a south German from the same region as Konrad Kyeser. His work, published about 1430, contains the first "certain representation" (Bertrand Gille) of the crank-and-connecting-rod system with flywheel,[55] in addition to hoisting apparatus, windmills, and a diver equipped with waterproof tunic, lead-soled shoes, and diving helmet. The author's consistent practicality, and the appearance of similar diver's rigs in other manuscripts, suggest that such gear was actually used in recovering sunken cargo.[56]

Paolo Toscanelli (1397–1482), a Florentine physician, geographer, and astronomer. To Brunelleschi's dome, Toscanelli added a great gnomon, which with the marble-flagstoned cathedral floor formed a giant sundial, determining the summer solstice and the dates of movable feasts. Toscanelli taught mathematics to Leonardo and gave impetus to Columbus; having compiled a map to show how the earth could be circumnavigated, he speculated on a westward route to India in a letter to an adviser of Alfonso V in Lisbon, who showed it to Columbus.[57]

Nicholas of Cusa (1401–1464), cardinal, Church reformer, mathematician, and experimental scientist, who like Nicholas Oresme believed that the earth rotates on its axis every twenty-four hours. He too expressed skepticism of received wisdom: "All human knowledge is mere conjecture and man's wisdom is to recognize his ignorance" (*De docta ignorantia,* On learned ignorance, 1440).[58] His study of plant growth proved that plants take nourishment from the air, and that the air has weight. He also discovered a dozen lost comedies of the Roman playwright Plautus.

Leon Battista Alberti (1404–1472), sometimes described as the prototype of the Renaissance man. His range of talents included sculpture, poetry, mathematics, and cryptology; his fame rests chiefly on three classic studies: the dialogue treatise *Della famiglia* (On the family), the book *Della pittura* (On painting), and especially the ten-volume *De re aedificatoria* (On architecture), printed in 1485, covering a range of subject matter but centering on town planning. "Curious, greedy for knowledge, endeavoring to understand, to explain, and to generalize" (Bertrand Gille), he has often been compared with Leonardo in his breadth of outlook.[59]

Roberto Valturio (b. 1413), whose *De re militari* (On military matters) became in 1472 the first engineering work to be printed. A copy was in Leonardo's possession. Among Valturio's devices was a windmill-propelled tank similar to that of Guido da Vigevano, and a boat propelled by five pairs of paddle wheels cranked from a single power source.[60]

Among Leonardo's own generation, some names are especially worthy of note:

Johannes Müller (1436–1476), better known as Regiomon-

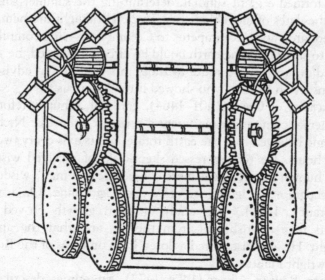

Windmill-propelled tank, sketched by Roberto Valturio. [Science Museum, London.]

tanus, who along with Fra Luca Pacioli is credited with stimulating critical examination of Ptolemaic astronomy. Regiomontanus (Latin for Müller's birthplace, Königsberg), in collaboration with Austrian mathematician Georg Feuerbach (1423–1461), produced new translations of Ptolemy's works that ultimately led to the revolution in cosmology which produced the Copernican system.[61]

Francesco di Giorgio Martini (1439–1502), another Sienese, a painter, sculptor, city planner, and architect, besides being an engineer whose talent "equaled that of Leonardo" (Bertrand Gille).[62] He designed fortresses and weapons, including an ancestor of the land mine. A copy of his manuscript Trattato di architettura civile e militare (Treatise on civil and military architecture) survives with Leonardo's annotations.[63] Among its devices is a more sophisticated version of Villard de Honnecourt's waterpowered saw: a crank and connecting rod move the saw, with a device to advance the workpiece.[64] Even more notable is the assembly of suspended weights on short arms that anticipates by 300 years one of Watt's most elegant inventions, the fly-ball governor. Another of Francesco's sketches shows an early version of the water turbine, an improved horizontal waterwheel driven by a stream directed on it by a conduit.[65]

Fra Luca Pacioli (1450–1520), a mathematician closely associated with Leonardo. He composed the Summa de arithmetica, geometria, proportioni et proportionalità (Synthesis of arithmetic, geometry, proportions, and proportionality), published in Venice in 1494. Based on the work of Leonardo Fibonacci, it was the second printed textbook on mathematics (the Treviso Arithmetic of 1478 was the first) and contains a pioneering treatise on double-entry bookkeeping. A second book, De divina proportione (On divine proportion), was illustrated by Leonardo da Vinci's drawings of symmetrical figures.[66]

Polydore Vergil (1470–1555), whose De rerum inventoribus (On the inventors of things, 1499) was the world's first history of technology.[67] Polydore inaugurated a long-lasting tradition of Eurocentrism in technology history; of Asia's contributions he was aware only of cotton and silk. Sent to England by the pope to help collect Peter's pence, a special English contribution to Rome, he stayed long enough to write a three-volume history of England's recent kings that became a principal source for Shakespeare's historical plays.

Marc Antonio della Torre (1473–1513), a professor at the University of Padua and close friend of Leonardo. His early death prevented fruition of an important project, an anatomical textbook planned with Leonardo; had they found time to execute it, Charles Singer speculates, "the progress of anatomy and physiology would have been advanced by centuries."[68]

Thus Leonardo had numerous precursors, peers, and associates in the creation of his "notebooks"—in Ivor Hart's description, "thousands of pages . . . [the] fevered and disordered activity of a lifetime—notes that teemed with scientific discussions based on observations and experiments; notes that swept through a wide range of problems in art, science, philosophy, and engineering."[69] Coming into the world as an illegitimate son, he missed out on university education, but he nevertheless had an education, and not a bad one. In the studio of Andrea del Verrocchio, besides painting and sculpture, he learned something of anatomy and algorism, how to cast guns, bells, and statues, and the mechanical arts, "a smattering of everything," which he rapidly expanded through "an immense and attentive curiosity" (Bertrand Gille).[70]

One of the first, if not the first, to study intensively the problem of friction in machine components, Leonardo sketched ball bearings and roller bearings that were new to him and probably to his time, although bearings are believed to have been used by the Chinese, the Romans, and other ancients. To the problem of translating rotary into reciprocal motion and vice versa he offered several solutions involving skillfully designed gears, one of which was employed by "all the constructors of machines in the sixteenth century" (Bertrand Gille).[71]

Where his conceptions represent improvements on existing devices they are often impressive, like the file-making machine

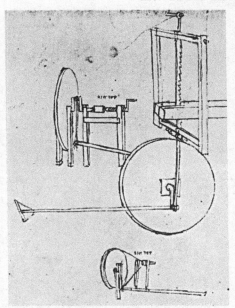

*Lathe and mechanical saw, drawn by
Leonardo da Vinci. [From Codex Atlanticus, 381 r.b. Science Museum, London.]*

whose hammer, mounted on a movable platform, is advanced
by a long screw; a gear system coordinates the rise and fall of
the hammer with the turn of the screw while advancing the
file the proper distance for each notch.[72] Also in the realm of
mechanical engineering, the notebooks include a lathe, hand
powered or treadle powered with flywheel, a mechanical saw, a
screw-cutting machine, a variable-speed gearing device, a
mechanical turnspit, a lens grinder in which grindstone and
lens revolve at different speeds,[73] a log-boring machine to meet
the demand for city water conduits, a variety of trip-hammers,
and many more variations on known themes.[74] Among his conceptions in textile machinery is a teaseling machine that
stretched the cloth between two rollers, one of which, turned
by a horse-powered winch, pulled the cloth under a beam
armed with teasel heads; a machine much like it, introduced in

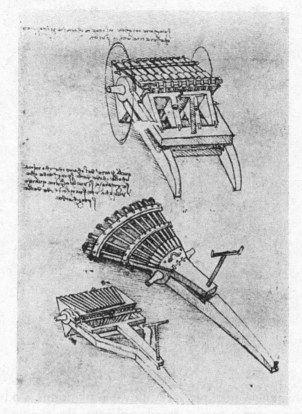

Leonardo da Vinci: a rack of guns that rotated for firing, cooling, and loading. [From Codex Atlanticus, 56 v.a. Science Museum, London.]

the British textile industry, caused riots among seventeenth-century hand teaselers.[75] A multiple-spindle spinning machine of Leonardo's was likewise finally realized in Britain in the seventeenth century.[76]

In military engineering, Leonardo improved on Konrad Kyeser's turntable battery with a triple rack of guns that rotated into positions for firing, cooling, and loading, and on the wind-propelled chariots of Guido da Vigevano and Roberto Valturio with an armored wagon hand-cranked from inside. His breech-loading cannon have a modern look, as

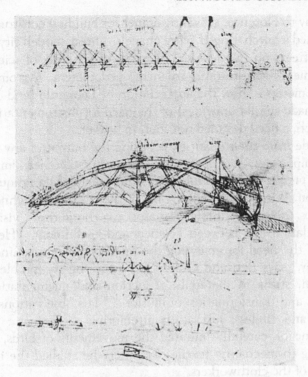

Leonardo da Vinci: truss bridge above, swing bridge center, and pontoon bridge (barely visible) below.
[From Codex Atlanticus, 312 r.a., Science Museum, London.]

does his explosive shell, which he describes with innocent enthusiasm as "the most deadly machine that exists." His interest in the paths of projectiles foreshadows the study of ballistics, an important contributor to the cooperation of science and technology. For a fortified island, where a bascule (draw) bridge would be inappropriate, he designed a swing bridge, mounted on a pivot—an idea that may have been wholly original and that, like the bascule, was resurrected and built many times by modern bridge engineers.[77]

In civil engineering, Leonardo produced designs for truss bridges, plans for making the Arno navigable by ocean ships all the way to Florence, swiveling cranes for building construction, a domed church ringed with chapels, and a model city plan with streets on two levels.[78] Sketches of an array of scientific instruments—hydrostatic balances, pedometers, hygrometers, anemometers—show the appreciation by Leonardo (and others who made similar drawings) of the need for instrumentation in research, a need destined not soon to be met.

Aside from their quantity, eclipsing the output of any of his contemporaries, the quality of Leonardo's sketches, "miraculously precise and graceful," gives the notebooks a unique distinction. But from the viewpoint of the history of technology, perhaps their most interesting aspect is the author's vision of the relationship between science and technology. "He perceived the need for analyzing the more complex machinery of the day," says Bertrand Gille, "and attacking the problems of friction, stress in materials, reduction and augmentation of power, and transformation of motion."[79] His observations were acute and tireless; just as in attempting to design a flying machine he carefully imitated the movements of birds, so in striving to mechanize textile machinery he studied the movements of the clothworkers.

Unfortunately, through a series of accidents Leonardo's notebooks remained unpublished and almost unknown until centuries after his death; some of the most important were rediscovered in a Madrid archive as late as 1967. Consequently, most of his conceptions never bore fruit. Yet relatively little was lost. The work of Leonardo's colleagues, from whom he himself had freely borrowed, continued without interruption to be copied, adapted, and, by later ethical standards, plagiarized. Francesco di Giorgio Martini's sketches of a roller mill, a horse treadmill, and a suction pump were copied without attribution in sixteenth- and seventeenth-century books and, passing to the East, were incorporated into the great Chinese encyclopedia of 1726, symbolizing the transformation that had taken place in the technological relationship of East and West.[80]

Thus, even if the specific ideas depicted in Leonardo's notebooks had little impact, the spirit that created them, "the irrepressible taste for mechanical achievements" shared by Leonardo and his fellow artist-engineers, their "constant and generalized preoccupation with machines and mechanical solutions" (Carlo Cipolla) had tremendous influence.[81] True, "the engineers' drawings [were] sometimes more advanced than their practical achievements" (Bertrand Gille),[82] but their conceptual renderings were often legitimate auguries of the future (even Leonardo's bravura conception of a bridge over the Golden Horn at Constantinople has been realized in the twentieth century). The quantity and quality of their ideas "self-reinforced" (Cipolla); the stream of books on mechanics became, in the two centuries that followed, and with the aid of the printing press, a torrent.

Fifteenth-century Technology: Incremental Gains

The innovations actually introduced into the technology of the fifteenth century stand in contrast to the freewheeling ideas of the artist-engineers. Where the drawings in the sketchbooks soar beyond the existing means of realization, the changes introduced in the forges, workshops, and mines were nearly all small, practical, and incremental. Some were among the more down-to-earth of the inspirations of the artist-engineers, some were products of more obscure working engineers, and some were contributions of anonymous smiths, masons, and craftsmen. Wider diffusion of many devices is reflected in their familiar treatment in iconography, such as the carpenter's brace and bit (crank application) shown in a basket carried by a Roman soldier in Meister Francke's *Carrying the Cross* (1424).[83] Taken all together, the incremental improvements, newfound applications, and wider diffusion added up to significant advance through the century and pointed the way to the future no less than did the imaginative renderings of Leonardo and his peers.

The turbine, a conception of Francesco di Giorgio, was actually put to work, though for an unexpected purpose. An advanced version of the waterwheel, deriving power from water (or gas) passing through it to spin an outer runner (rotor) armed with blades, the turbine eventually powered steamships and electric generators, but its fifteenth-century function was to serve as a turnspit governor; the hotter the fire burned, the faster the hot gas spun the turbine above and turned the roast.[84]

Leonardo proposed employing the turbine principle in a centrifugal pump to create a vortex high enough to spill over the containing vessel, as a means of draining swamps. By this time an anonymous inventor, probably in the Low Countries, had come up with a more practical idea: a radically improved windmill design. The tower, or hollow-post, mill mounted the mill mechanism in a revolving turret that could be turned without

In this version of a paddleboat sketched by Mariano di Jacopo Taccola, the current turns the paddle wheels, which reel in the rope, propelling the boat upstream (the man helps by pulling on the rope). [Bibliothèque Nationale, Ms. lat. 7239, f. 87.]

needing to move the entire structure. The water was lifted either by the scoop action of a vertical wheel or by an Archimedes' screw.[85]

The other great prime mover, the waterwheel, continued to expand its functions, creating lakes and streams (and sometimes impeding navigation) while powering industrial operations that now included smelting, forging, cutting, shaping, grinding, and polishing metals. It helped produce beer, olive oil, mustard, paper, coins, wire, and silk; it lent its powerful assistance to fulling cloth, sawing wood, boring pipes, and (by around 1500) ventilating mines. It supplied the power for an improvement in city water supply that began in south Germany with the introduction of piston pumps driven by undershot waterwheels.[86]

The Chinese treadmill-paddle-wheel boat, an application of the waterwheel in reverse, either reached Europe in the fifteenth century by stimulus diffusion (as Needham believes) or was independently invented. It appears in several manuscripts of the artist-engineers and in the following century was built in Spain, where it was long used for harbor transport.[87]

One of waterpower's most important fifteenth-century applications probably came in the task of pumping out mine shafts.[88]

Cutaway model of man-powered paddle-wheel boat, after a sketch by Leonardo da Vinci, in the Museo della Scienze e della Tecnica, Milan.

Of the variety of pumps, bucket chains, animal treadmills, and windlass-powered devices shown in Agricola's classic *De re metallica* (On metallic matters), published in 1556, some were certainly in operation in the fifteenth century. A famous one was designed by Jacob Thurzo of Cracow to deal with the chronic water inrushes of the deep silver-lead mines of the Carpathians in Hungary: an endless two-drum bucket chain powered by an animal treadmill. The device became the technical basis for a major new mining enterprise of Jacob Fugger, the moneyman of Augsburg, financial backer of Maximilian of Habsburg.[89]

Another anonymous innovation that appeared in the metal mines was the wagon mounted on wooden rails, drawn, until the arrival of the steam engine, by animal power. The fifteenth-century carriage makers supplied the future railroad with the pivoted front axle, ancestor of the bogie.[90]

In 1451 in the Austrian Tyrol, Johannes Funcken invented a new smelting technique to separate silver from lead, an improved version of the cupellation process used by the Romans and described by Theophilus Presbyter. It involved heating cakes of lead, copper, and silver to run off into ladles in which an experienced smelter could separate out the silver. The process supplied another technical assist to Jacob Fugger's mining and metallurgical enterprise and to the Habsburg hegemony in central Europe.[91] The even better mercury amalgam process arrived in time to aid in the Spanish exploitation of American silver mines, also to the benefit of the Habsburgs.

As the blast furnace and refinery delivered an increasing supply of iron to the forge, the smith received help in handling it from the waterwheel via a new device (old in China), the tilt hammer, or trip-hammer. A heavy iron head on a wooden shaft was lifted and released by a drum armed with cams. Rising, it struck a wooden spring beam; the spring's recoil added force to the downstroke. Alternatively an iron block in the floor under the hammer tail achieved the same result.[92]

Around 1500, and probably before, the slitting mill made its appearance in the iron-rich Liège district. The demand for nails was increasing; the new mechanism provided the smith with slender rods easily converted into nails. It consisted basically of a pair of rotary disk cutters turning in opposite directions. "The first piece of true machinery after the power hammer to be introduced . . . of even greater importance, it contained the elements of the rolling mill" (W. K. V. Gale). In fact, it gave birth to the rolling mill: two iron cylinders powered by water-wheel flattening a bar of iron passed between them.[93] Into modern times, the rolling mill remained a basic tool of the iron and steel industry.

In the year 1500 iron production for the whole of Europe amounted to the impressive figure of 60,000 tons, according to the estimate of Rupert Hall. Nor was iron the only metal experiencing a boom. Church bells and cannon created a demand for bronze that stimulated larger and larger installations. Both bells and cannon required large quantities of metal and a high degree of skill; bell metal, 23 to 25 percent tin, depended on precision in casting to ensure the proper ring.[94] Among Leonardo's sketches are a reverberatory furnace (one in which the ore is not in contact with the fuel) to produce large quantities of metal for bell founding and cannon casting, and a crucible furnace in which six crucible pots were aligned in a sloping flue up which the flame swept "like a blowtorch."[95]

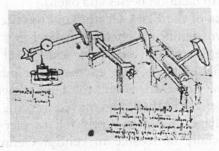

Tilt hammers, sketched by Leonardo.
[From Codex Atlanticus, 21 r.a. Science Museum, London.]

Leonardo's "furnace of the controlled flame," in which six crucible pots were aligned in a sloping flue. Model in the Museo della Scienze e della Tecnica, Milan.

The expansion of mining and metallurgy benefited agriculture by increasing the number of tools available and reducing their cost. Metal implements of every sort "figured a great deal more commonly in the everyday life of the sixteenth century than in that of the fourteenth."[96] A peasant anywhere in Europe had a far better chance than his great-grandfather of owning not only the basic complement of farm tools but plow and cart (not to mention horses and oxen).

In the wake of the Black Death, land too became cheaper, leading to two significant changes in land use. One was the shift to sheep farming, especially widespread in England, where open-field villages were enclosed, the inhabitants sent packing by the new owners, and the land turned over to the sheep. Shepherds and their dogs moved into the manor house while the village dwellings slowly fell to pieces. A second change, especially noticeable in central Germany, was the regeneration of forest where clearings were abandoned for lack of labor to cultivate them. Scrub birch and hazel took over the empty fields, to be replaced in time by beech and other tall forest trees. Georges Duby writes, "The return of natural vegetation in the fourteenth and fifteenth centuries is an episode . . . of equal importance to the adventure of clearing the wastes."[97]

The enclosure movement reflected the continued growth of the textile industry, which in the course of the fifteenth century improved both its spinning and its weaving instruments and introduced changes in the organization of work. New patterns of consumption developed, and geographical shifts in manufacture took place.[98]

An attachment to the spinning wheel, the flyer spindle, sketched by Leonardo and once believed to have been invented by him, is now known to have been borrowed from the silk-throwing process and long used in spinning wool in northern Italy. In this device, as the spindle turned, a flyer—a fork with toothed projections—revolved around it at a different speed, giving the yarn an extra twist. Fork and spindle were moved by separate wheels, powered by the same transmission belt but turning at different rates because of their different diameters.[99]

The final medieval improvement to the spinning wheel, probably occurring late in the fifteenth century, was the addition of a treadle to power the wheel, leaving the spinner's right hand free to regulate the delicate task of feeding the raw material to the spindle. The result was a product that was more regular and of better quality.[100]

In the organization of work, while the putting-out system survived in eastern Europe, it was gradually replaced in the West by the beginnings of a true factory system. In some places the factory was partly dispersed; in Florence a cadre of finishers was employed in the merchant's shop while the weavers continued to work in their own homes. In both home and workshop, bells rang for the beginning and end of the working day as well as for meals, and inspectors regularly monitored all the workers.[101]

In a second form of factory that appeared in England, the sheep-raising landowner established production on his own manor, outside the jurisdiction of both city and guild regulations. At first, as in Florence, the workers labored in their homes and were visited by inspectors. Later they sometimes operated in a central workshop such as the one described, factually but with considerable exaggeration, in the sixteenth-century ballad "The Pleasant History of John Winchcomb, Called Jack of Newbury." The establishment of Jack of Newbury (d. 1519), according to the ballad, was equipped with a thousand looms and employed over a thousand persons, all under one roof. More reliable accounts describe similar but more modest arrangements. The historian John Leland (c. 1506–1552) tells of an entrepreneur who installed his textile factory in an abbey, in which "every corner of the vast houses . . . be full of looms." Later the same man acquired another monastery, laying out streets around it, each dedicated to a special function of cloth production.[102]

As wages rose and prices fell in the aftermath of the Black Death, patterns of consumption shifted. The wool and silk industries found it profitable to produce less expensive fabrics to appeal to middle-class customers, while cotton and linen manufacturers exploited the lower end of the market, providing peasant households with bedding, table linen, and undergarments to take the place of their traditional homespun. To serve the new customers, cloth-making centers specializing in cheaper grades sprang up in England, Holland, Germany, France, Spain, and Switzerland.[103]

The northern Italian cotton industry received competition from southern Germany in the form of cheap but durable fustian, a mixture of linen and cotton known in Europe for at least three centuries but now mass-produced. The German fabric was woven with a warp of local linen fibers and a weft of cotton yarn imported from Venice and Milan. Later the weft was also produced locally, spun from bales of raw cotton bought in northern Italian ports and transported over the Alps. The German fustian industry, which founded several fortunes including that of the Fuggers, was decentralized and largely rural. Merchant capitalists in the towns employed numbers of weavers in the countryside, who enlisted local spinstresses to spin the cotton weft thread and bought linen warp in the local market. The weavers were "less skilled, less supervised, and more poorly paid" than those of the Italian cities, accounting for the lower quality and price of their product, which gradually undermined the Italian cotton industry. A shift completed in the sixteenth century moved cotton from the Mediterranean ports and Italian cities to the Atlantic ports and northern Europe, with raw material beginning to come from the New World.[104]

Another, quite different, ancestor of modern mass production appeared in Venice. In the Venetian Arsenal the arming and equipping of war galleys was accomplished by a primitive form of the assembly line. A Spanish visitor, Pero Tafur, wrote an account of the operation that he observed in 1436:

> As one enters the gate there is a great street on either hand with the sea in the middle, and on one side are windows opening out of the houses of the arsenal, and the same on the other side, and out came a galley towed by a boat, and from the windows they handed out . . . from one the cordage, from another the bread, from another the arms, and from another the ballistas and mortars, and so . . . everything that was required, and when the galley had reached the end of the street . . . she was equipped from end to end. In this manner there came out ten galleys, fully armed, between the hours of three and nine.[105]

In civil engineering, Renaissance architecture created a building boom whose principal technical advances came in lifting machinery, such as the counterweighted pulley hoist devised by Brunelleschi which allowed a rope drum to reverse and set down a load without disturbing the motion of the winch or animal treadmill and which delivered stone blocks, brick, lime, sand, and water for the cupola of the Florentine Duomo. Two pinions, an upper and a lower, could be made to connect with a large wheel, lifting the load or the counterweight.[106]

The long-standing problem of lock gates was mastered by an idea from China, given perfection by Leonardo. Early lock gates, either simple double doors swinging on hinges or vertical portcullises that lifted straight up, offered insufficient resistance to the pressure of water on the upstream side of the lock basin, and the portcullis type had the disadvantage of needing a high clearance for boats. The Chinese design that reached the West in the fourteenth century provided double doors facing upstream at an obtuse angle, so that the pressure of the stream only forced them more tightly together. Leonardo, after

System of portcullis lock gates, sketched by Francesco di Giorgio. [Trattato dell'architettura, Codex Ashburnham, 361 f., 41a. Biblioteca Medicea-Laurenziana, Florence.]

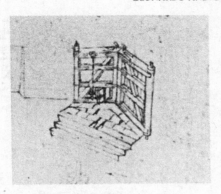

Leonardo's mitered canal lock-gates.
[From Codex Atlanticus, 240 r.c.
Science Museum, London.]

studying the locks in the Visconti canal system around Milan, added a stylish touch: mitered gate edges that met in a snug fit. Water was admitted to the lock basin through small sluices cut in the gate.[107]

The mechanical clock spread rapidly in the fifteenth century, becoming a feature of private houses as well as royal palaces and communal towers.[108] A late-fifteenth-century invention, the mainspring, second in importance only to the escapement, made timepieces not only portable but cheap.[109] The pocket version got its name "watch" from the town watchmen who took to carrying it.[110] But although the watches had alarms and struck the hours, they kept only indifferent time. The trouble lay in the variable torque of the mainspring, which grew weaker as it unwound. The solution lay in the fusee, a device sketched by Konrad Kyeser in *Bellifortis* as part of a crossbow and applied to clockwork in Prague in 1424: it was a cone around which a cord connected to the spring was wound. As the spring uncoiled, the diameter of the cone increased, augmenting the leverage and compensating for the weakening of the spring's pull.[111]

The earliest application of the mechanical clock to scientific use came in 1484 when Walterus, landgrave of Hesse, another prince with scientific and technological interests, measured the

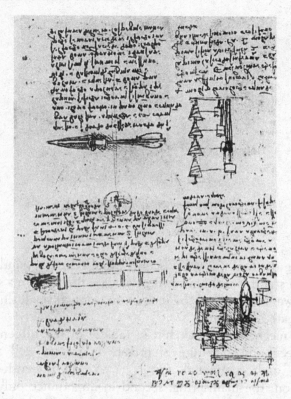

Fusees and clock mechanism, on right, sketched by Leonardo (on the upper left, a finned explosive projectile). [M. B, f 50v. Science Museum, London.]

interval between the transits of the sun from noon to noon, using a mechanical clock.[112]

In the decorative arts, two landmark technical innovations appeared. Oil as a painting medium was mentioned by Theophilus Presbyter in the twelfth century, but egg-based tempera reigned supreme until a process for refining linseed oil, producing volatile solvents, was developed, mainly in Venice. Pigments dispersed in the treated oil created a responsive medium, exploited early by the Van Eyck brothers in Bruges and a number of artists in Italy (and seized on by Gutenberg for printer's ink).

The maturing of the casting art encouraged the creation of bronze statuary; the first equestrian figure to grace a public square, a statue of the Florentine condottiere Erasmo Gattamelata, was executed in 1453 by Donatello, who also cast a statue of David for the Medici palace, the first bronze fountain piece of the modern world.[113]

Thus medieval technology made a direct contribution to art. It made a larger indirect contribution in helping to create such fortunes as that of the Medici. Besides Donatello's statue of David, Cosimo de' Medici (1389–1464) commissioned several madonnas by Fra Lippo Lippi, frescoes by Fra Angelico (for the monastery of San Marco), one of the first great equestrian frescoes, by Andrea del Castagno, a madonna by Flemish master Rogier van der Weyden, terra-cotta reliefs depicting the labors of the field, by Luca Della Robbia, and for his private chapel a Procession of the Magi that included portraits of members of the Medici family, by Benozzo Gozzoli.[114]

The Ocean Ship

European seaborne commerce expanded in every dimension in the fifteenth century: more ships, larger tonnages, better port facilities. Quayside loading and unloading of sailing ships was now established in northern, southern, and Atlantic ports. The Low Countries pioneered technology for harbor maintenance, such as the dredge built by the Dutch to scrape the harbor bottom at Middelburg with a ponderous rake, loosening silt to be carried out by tidal current. Leonardo sketched a more sophisticated solution in the form of a twin-hulled dredge with scoops mounted on a vertical drum, but effective instruments awaited the next century.[115]

By far the most important new element in navigation was the full-rigged ship, "the great invention of European ship designers in the Middle Ages" (Richard Unger),[116] which "enabled Europeans to harness the energy of the wind over the seas to an extent inconceivable to previous times" (Carlo Cipolla).[117] Its principal fifteenth-century form, the carrack, represented the final step in the centuries-long evolution of the round ship: essentially the northern cog, as modified by Mediterranean builders, with further refinements added by Basque shipbuilders of the Bay of Biscay. A large, heavy tub with a big spread of canvas, the carrack had a stout length-to-breadth ratio of three and a half to one or less. The massive skeleton ribs that framed its hull, now carvel-built in northern as in southern yards, supported two or even three decks. A majestic sterncastle rose aft of the mainmast, balanced by a smaller but higher forecastle.[118] Its edge-to-edge planking was tightly caulked with oakum (shredded hemp) and tar or pitch

Model of Flemish carrack, c. 1480, with lateen sail on mizzen mast, and stern rudder. [Science Museum, London.]

and given an outer protection of wales and skids to cushion the collision with the quay.[119] Few hatches and no companionway helped make it watertight in heavy weather.[120] The tiller that operated its sternpost rudder passed through a port in the stern to a whipstaff.

Of its three masts, the main and foremast were square-rigged and supplied most of the power. The mizzen, rising from the sterncastle, was lateen, for control. The huge mainsail hung from a yard as long as the ship itself, below a much smaller topsail; the foremast carried a single square sail. By the end of the century another small sail, the spritsail, on the bowsprit, assisted the lateen in control.[121] Genoa and Marseilles were reputed sources of the best sailcloth (cotton or linen canvas). The square sails were now easier to handle, thanks to improvements in the ropes. The mainsail could even be used to assist the tacking maneuver; as the ship came into the wind, it was raised momentarily to swing the bow over to the new tack.[122] The multiplicity of sails proved invaluable when it came to navigating narrow waters, and did not demand more crew, since the sails were worked one at a time.[123]

The best bulk carrier yet built anywhere, the carrack could take up to a thousand tons of wheat, salt, and timber in its capacious hold.[124] Ranging freely and securely from the Baltic to the eastern Mediterranean, entirely supplanting the sail-and-oar galley on the Italy to Flanders run, it supplied the critical means for implementing the new interdependence of the economies of northern and southern Europe.

Columbus's *Santa Maria* was a carrack, though one of quite modest proportions, probably not much more than a hundred tons. His two smaller ships, the *Niña* and *Pinta*, were products of a second, parallel line of development that began about 1440.[125] The caravel was a shipbuilder's solution to a very specific navigation problem: that encountered by Portuguese mariners groping their way down the west coast of Africa in search of the passage eastward to Asia. Carrying mixed or lateen rig and weighted with a cargo of no more than fifty tons, the slim caravel (the name a reminder of its carvel construction) had excellent sailing characteristics, including an ability to sail close to the wind that greatly facilitated the return voyage north to Portugal. Before the wind it was capable of a speed of up to eleven knots. Columbus's *Niña* and *Pinta*, returning from America in 1493, made a day's run of 198 miles.[126] The caravel's small crew and minimum supply requirements suited it to exploration of unknown and distant waters, and its maneuverability allowed it to fight off a lee shore even better than could the carrack.

The magnetic compass was now a mature navigation instrument. The fact that the needle did not point exactly north had been duly noted and allowed for; since no one knew why it pointed north in the first place, the discovery made little difference.[127] Simplified versions of the astrolabe and its variant, the quadrant, measured the angles of the two Guardians in relation to the North Star; the resulting data used in conjunction with tables gave latitude within about twenty-five miles.[128] As the Portuguese African ventures reached further and further south, they proved the earth's sphericity beyond a cavil by sighting new constellations, including the spectacular Southern Cross, but lost their ancient guiding light, the North Star. In 1484 King John II appointed a commission of mathematicians to study the problem and draw up tables of declination of the sun to be used at sea in conjunction with the astrolabe or quadrant; by determining the sun's height at midday and consulting the tables, sailors could ascertain latitude.[129] A new navigation technique was born: the skipper first sought the correct latitude for a certain port or point of land, then ran along the line of latitude to his target destination.[130] To the tables of declination were added charts of known coasts and pilotage information. Arab and Chinese pilots of the Indian Ocean already knew how to find latitude, but they never adopted the European custom of carrying charts on board that made it easy to repeat an exploratory voyage with high accuracy.

Despite the advances, navigation at the end of the century still demanded much in the way of experience, judgment, and instinct. Liberation from dead reckoning required a means of determining longitude, which awaited the invention of the chronometer in the eighteenth century. The advances of the fifteenth made it possible, not easy or safe, to explore the immense Ocean Sea and its rumored, but unnamed and uncharted, coasts and islands. But what technology makes possible, someone undertakes.

General knowledge of geography was expanded by publication in Latin in 1406 of the tardily translated *Guide to Geography* of Ptolemy, who had compiled his gazetteer-atlas-world-map information in the second century A.D. Necessarily sketchy and inaccurate, it nevertheless added considerable detail to medieval knowledge. Ptolemy's errors did not all go undetected; Pope Pius II (reigned 1458–1464) exposed one, that of an Indian Ocean landlocked by a dim southern continent. Most significantly, while strengthening the perception of a spherical earth, Ptolemy perpetuated the optimistic reduction of its size and proportion of water made by Marco Polo and Pierre d'Ailly. Columbus, who studied all three authorities, and was prejudiced to begin with, inevitably accepted their calculations. The oldest extant map in the form of a globe, made by Martin Behaim, a German who had been long resident at the Portuguese court, was of equally little help. Columbus took a Behaim globe with him aboard the *Santa Maria*, but it gave him false reassurance on the size of the oceans and was even curiously out of date on details of the coast of Africa.

Misconceptions still linger about the "spice trade" that motivated the voyages of discovery. Europeans have been credited with an insatiable appetite for seasonings, attributed either to monotony of diet or to a need to disguise the taste of meat that was thought to have chronically spoiled owing to lack of refrigeration. Both notions contain only a particle of truth. The peasant diet was certainly monotonous, but peasants could not afford imported spices. The diet of the well-to-do, the customers for spices, was a different matter: meat and fish at all seasons, and a list of fruits and vegetables that increased through the Middle Ages (oranges and lemons early, lettuce in the fourteenth century, artichokes and cantaloupe in the fifteenth), even before the influx of new products from the Americas. The popular *Tacuinum sanitatis* (Health handbook), which appeared in many fourteenth-century versions, also mentions among its vegetables and fruits spinach, asparagus, leeks, turnips, pomegranates, watermelons, cucumbers, green squash, sour cherries, cabbage, beets, and chestnuts.[131]

In respect to food preservation, medieval Europe was hardly worse off than nineteenth- and early-twentieth-century Europe and America. A number of techniques were available: smoking, salting, and drying, and also on-the-hoof preservation—oxen, goats, sheep, poultry, and pigs driven live to city markets, game and domestic animals killed and eaten forthwith. Affluent households in town and country stocked fish tanks and ponds. There was little need to disguise spoiled meat (which the wealthy would have refused to eat). There was, however, a need to add flavor to meat that was usually subjected to long cooking in liquid because of its toughness.

That fact alone hardly explains the importance of the spice trade. The real keys to the mystery are two. The first is the physical character of spices: extreme compactness in proportion to value, and resistance to spoilage. Though not outrageously expensive to the consumer—a little pepper or saffron goes a long way—they carried very high price tags for the amount of cargo space they took up. This was a consideration of overriding importance when most ships could carry no more than a hundred or two hundred tons. A merchant reserving cargo space aboard a Venetian or Genoese vessel on the Syria run filled his quota with the most valuable merchandise per weight that he could find: gold and silver ornaments, jewelry, silk, and spices.

The second key to the mystery is a matter of vocabulary. The medieval use of the word "spices" covered a vast multitude of useful commodities, only a portion of which were destined for the cooking pot. Robert Lopez summarizes them as "seasonings, perfumes, dyestuffs, and medicinals."[132] Florentine merchant Francesco Balducci Pegolotti's *La pratica della mercatura* (The practice of commerce), in its comprehensive list of 288 "spices" carried in fourteenth-century commerce, enumerates, alongside anise, cinnamon, cumin, ginger, cloves, nutmeg, pepper, sugar, fennel, and citron, such pharmaceuticals, dyes, industrial additives, and miscellaneous items as camphor, wax, alum, rosewater, cotton thread, paper of Damascus, glue, ivory, indigo, frankincense, shellac, musk, linseed oil, niter salt, soda ash, soap, turpentine, Venetian copper, nux vomica, and gold leaf.[133] This large array originated in a wide scattering of sources in India, Indonesia, and southeast Asia, and moved to Europe partly by ship (via the Persian Gulf or Red Sea), partly by caravan or pack train, with many transshipments, many tolls, and much danger of loss. A sea route that would permit a ship to sail from Europe all the way to the "Indies," load up, and sail back with a hold full of "spices" would guarantee a fortune per voyage.

The spice trade did not begin in the Middle Ages. Pliny comments on the widespread use of pepper, with which Rome was so plentifully supplied and with which the barbarian Goths were so familiar that when Alaric exacted a ransom from the city in 408 he included in his demands 3,000 pounds of pepper.[134] Nor did the spices by themselves account for the Age of Exploration. Other motives entered in. Religious proselytizing was as old as Christianity and had won converts, willing or reluctant, among third-century Goths, fifth-century Franks, the wild Vikings of Scandinavia, and the Poles and Magyars of eastern Europe. Whether proselytizing came first and profit second or vice versa may be an open question, but to the lure of spices should be added that of certain other merchandise, notably gold, increasingly needed to fuel the Commercial Revolution. Spanish conquistador Bernal Díaz voiced a Christian-capitalist ideal in expressing the wish "to serve God and his Majesty, to give light to those who were in darkness, and to grow rich as all men desire to do." A modern historian has evaluated with succinct cynicism: "Religion supplied the pretext and gold the motive" (Cipolla).[135]

The medieval character of exploration and its motives is underlined by the kind of inducements offered to the explorers.

The Portuguese kings agreed to give their captains a share in newfound lands along with the profits of civil and criminal justice and the monopoly of mills, ovens, and salt. The letters patent given Venetian John Cabot by the king of England included the governorship of new lands, a monopoly on their produce, and duty-free importation, with a fifth of the profits to go to the Crown.[136]

Another motive was fishing. Cod disappeared from the coastal waters of Europe just as an improvement in fish-packing technology, a press to pack the salted cod into barrels, was invented. Basque and other fishermen may have found the fabulous codfish grounds of the Grand Banks before Cabot did in 1497–98 without ever advertising their discovery (as fishermen often do not).[137]

Finally there was the lure of the unknown but knowable, the opportunity the full-rigged ship gave to find answers to the mysteries that had baffled and fascinated European intellectuals from Ptolemy to Toscanelli. "One of the most powerful incentives for Atlantic exploration was the quest for islands" (J. R. S. Phillips) whose existence had been persistently bruited by sailors and mapmakers.[138]

That the lead was taken by the new and small kingdom of Portugal was owing partly to Portugal's unique geographical position, fronting the Atlantic but close to the gateway to the Mediterranean, and partly to the progress of the Reconquista. With the last Muslims driven out of Portugal, the natural continuation was to carry the war across the water to North Africa, which the Portuguese did in 1415 by seizing Ceuta, across the strait from Gibraltar. Quite apart from a route to the Indies, Africa itself was worthy of attention as a known source of gold bullion, and the early Portuguese exploration south was oriented to Africa's rather than Asia's wealth.[139]

Although recent scholarship has somewhat discounted the individual contribution of Prince Henry the Navigator, he remains a remarkable figure, employing funds available to him as master of the Order of Christ to attract geographers and savants to Lisbon and to fit out expeditions. The first fruits were the Madeira Islands to the south, already discovered by the ubiquitous Genoese in the fourteenth century and rediscovered for Portugal in 1418; and the great chain of the Azores (1427–1431), a third of the way across the Atlantic. Both archipelagos were barren of inhabitants, and both proved highly colonizable, congenial to the cultivation of sugar, one of the most treasured of the spices. The west coast of Africa was reconnoitered by a series of imaginative voyages in which the caravels turned the prevailing winds to advantage by first sailing well out into the Atlantic, then angling back to the African coast, where the few river mouths and inlets were one after another discovered. Trading with the natives netted gold, slaves, and elephant tusks. These last quickly captured the ivory market from the walrus tusks of Greenland, whose Viking colony, hard hit by the Black Death and a prolonged cold wave, gave up and retreated to Iceland. Greenland reverted to its Inuit (Eskimo) natives, thus putting an end to the sterile Scandinavian northern adventure just as the fruitful Portuguese southern adventure was picking up momentum.[140]

In both directions, west and south, distances proved disconcertingly longer than navigators had been led to believe by the authorities, but in 1488 the Cape of Good Hope was at last rounded, and in 1499 Vasco da Gama, a soldier given charge of a three-ship expedition, made it back to Lisbon with two ships loaded with enough spices to pay for the voyage several times over.

Da Gama's ocean trail was swiftly followed by ships of all the western European nations. Surprisingly, as Fernand Braudel has pointed out, it was not followed in reverse by Asian ships. Large Chinese multisailed and multidecked junks had shown themselves fully capable of long-distance ocean voyaging; Admiral Cheng Huo's fleet made a succession of voyages to India and East Africa between 1405 and 1433 Why the Chinese tamely abandoned the European spice traue to the Europeans remains a historical mystery.[141]

In the other direction, westward, Portuguese exploration was checkmated by its very success in discovering and colonizing the Azores. Using the Azores as a base, Portuguese mariners trying to sail into the teeth of prevailing winds got nowhere, but a southbound expedition, under Pedro Cabral, taking the usual long southwest tack followed by a return southeast, discovered Brazil. By that time, Columbus, still another Genoese who took service with Portugal, but ultimately sailed for Spain, had put into effect his own adventurous plan, which was to start not from the Azores but from the Canaries, a long-inhabited archipelago several hundred miles south, now in the possession of Spain. From there he was able to pick up favorable winds to carry him to what he imagined to be the islands and coasts of Asia, and to use the westerlies to get back to Spain.

One resounding irony of Columbus's voyage is that the New World produced none of the traditional spices he sought but supplied a trove of entirely new foodstuffs for the European table: maize (corn), potatoes, chocolate, peanuts, tomatoes, pineapples, green beans, lima beans, red and green peppers, tapioca, vanilla, and the turkey. At the same time, America gained many European crops: wheat, barley, broad beans, chick-peas (garbanzo beans), sugarcane. Asia and Africa were brought into the general exchange, Asia receiving sweet potatoes, pineapples, papaya melons, and chili peppers while giving America bananas, rice, and citrus fruits. Africa received maize, manioc, sweet potatoes, peanuts, and green beans, and sent to America yams, cowpeas, coconuts, coffee, and breadfruit.[142]

Yet another irony: Columbus's voyage, as it turned out, neither depended on nor demonstrated the sphericity of the earth, since he could have made the same trip, Spain to the West Indies and back, on a flat earth.

Whether Columbus was preceded by Irish missionaries, Bristol merchants, Basque fishermen, or anonymous Portuguese explorers was once regarded as worthy of scholarly debate. Today, as better acquaintance with medieval history improves our perspective on the Age of Exploration, it is easy to see that remarkable though Columbus's feat was, the European discovery of "America" was inevitable within a short time, and even without Cabral's fortuitous landfall in Brazil. Motives were sufficient, and means, developed over centuries, were ample.

Conjectural model of Columbus's Santa Maria. [Science Museum, London.]

1500 AND AFTER: "WESTERN CIVILIZATION"

The America that Columbus discovered for Europe had supported its human population for unknown thousands of years, long enough to develop its own civilizations which in many respects (for example, irrigation agriculture) were remarkable indeed. But its isolation from the rest of the world after the submerging of the Aleutian-Bering Sea crossing had imposed handicaps. The Americas offered no large animals suitable for riding and traction, although the Peruvians had domesticated the little llama for pack carrying. Maize was widely cultivated, but wheat was absent. In most regions tools remained of wood, bone, and stone. In the absence of traction animals, the wheel was not invented (except as a toy); consequently, the wheeled vehicle, the potter's wheel, the spinning wheel, and the water-wheel all remained unknown.

The discrepancy in technological levels conditioned much of the relationship between the European discoverer-adventurers and the native Americans. The first appalling consequences, however, were due to something else: the lack of immunity of the Americans to European diseases—influenza, malaria, measles, and above all smallpox. In Europe mainly a childhood disease that conferred immunity on adult survivors, in America smallpox lethally attacked people of all ages, creating steep, long-range demographic declines.[143] These catastrophes were by no means deliberately inflicted; on the contrary, the European invaders wanted a healthy and numerous native population for labor recruitment. Plenty of room for criticism of the treatment of the natives remains, not only, it should be noted, by the Spaniards, who have commonly been made the scapegoats, but by others, including the English and their American descendants. Lack of immunity, incidentally, cut both ways; Columbus's expedition has been credited with bringing back to Europe syphilis, a minor ailment in the Western Hemisphere but a ferocious one in Europe; yellow fever similarly discriminated against the explorer-invaders.[144]

No pathological disasters followed European expansion in Africa and Asia, which evidently shared immunities with their European neighbors. Not that Africa and Asia had reason to be happy with their visitations; Africa became a source of slaves to work the mining and agricultural enterprises of America, and Asia in due course felt the weight of imperialism. Yet the Europeanization of the world, whatever losses it has entailed along the way, is virtually complete and has been almost universally accepted at least in its material aspect. Few today want to return to the civilizations of Greece and Rome, or to those of the Aztecs and Incas.

Global Europeanization took several centuries and embraced much more than technology. But at the heart of the historical process that wrought the vast alteration lay the slow revolution in tools and processes that transpired in Mediterranean and northwest Europe between the sixth and sixteenth centuries. During that medieval millennium, Europe left the world of Rome far behind, while overtaking China and India. The rising technological level of medieval Europe is reflected in the improvement in daily life and work: from slave labor to free labor, from human drudgery to animal power and waterpower; from luxury handicrafts to mass production for mass markets; from handwritten manuscripts for a scattering of intellectuals to printed books for a large audience; in metal tools and metalware, profusion in place of scarcity; and a long list of useful novelties, from clocks to canal locks. And, not to overlook the dark side of progress, gunpowder weapons, the one legacy of medieval technology that was indeed "pernicious."

Europe built its new "Western civilization" on a material foundation that it created not merely by borrowing freely from others but by making its borrowings extraordinarily effective. "A technologically progressive society," says D. S. L. Cardwell, "is . . . both willing and able to accept and apply inventions from whatever source they may come."[145] Just how much of Europe's technology actually derived from Asia remains a mystery awaiting scholarly detective work. A comment of Joseph Needham in respect to cast iron may be more widely applied: "Admittedly there is no one clinching piece of evidence, but rather a mass of hints."[146] Asian priority in a wide range of innovations is established. Asia, however, showed little inclination to borrow, and so, after giving much to others, allowed its own technology to wither, as demonstrated in the history of the two epoch-making inventions of printing and firearms. Each originated in China, but each was allowed to languish, while Europe seized them in both hands to make them major instruments of change. An authority on technology transfer in the modern world asserts that the process "is not just a matter of moving some piece of hardware from one place to another . . . A material infrastructure is not enough. There must also be sufficient nonmaterial infrastructure."[147] In the "nonmaterial infrastructure" of medieval Europe was a spirit of progress whose ingredients included intellectual curiosity, a love of tinkering, an ambition "to serve God" and also "to grow rich as all men desire to do."

A sense of progress implies a sense of history, something missing among the Egyptians, Greeks, and Romans. "Lacking any objective understanding of the past—that is, lacking history," says Cardwell, "the hierarchical and slave-owning societies of classical antiquity failed to appreciate the great progress that had been achieved by and through technics."[148] On the contrary, the ancients were fond of looking back to what they conceived as a vanished "golden age," a conception the reverse of progress. The Christian Church, whose pioneering monastic orders made many practical and material contributions to medieval technology, also supplied a noncyclical, straight-line view of history that allowed scope for the idea of progress.

Optimistic and utilitarian, fifteenth-century Europe's craftsmen, smiths, engineers, and shipbuilders sought better ways to do things, make things, make things work. Carlo Cipolla identifies their keynote: "Machines came to play an increasingly important role in the production process."[149] The fifteenth century's Francesco di Giorgio Martini explained why: "Without mechanical ingenuity the strength of man is of small avail."[150] Francesco echoed Hugh of St. Victor's concept of man, "naked and unarmed" but equipped with reason in order to supplement his weak powers by invention.

In seeking ways to multiply the feeble strength of man, medieval Europe found its most effective instrument in the vertical waterwheel, the world's chief prime mover until the invention of the steam engine. Neither Rome nor China succeeded in harnessing its power to the extent that medieval Europe did. Terry Reynolds summarizes the encompassing role it achieved in the high Middle Ages: "The house medieval man lived in might have been made of wood sawed at a hydropowered sawmill . . . The flour he ate . . . the oil he put on his bread . . . the leather of the shoes he put on his feet and the textiles he wore on his back . . . the iron of his tools . . . the paper he wrote on" all were produced in part with the aid of waterpower.[151]

A valuable spin-off from the waterwheel was the encouragement it lent to experimentation with key mechanical auxiliaries: gears, cams, cranks, and flywheels. It did not at once bring on the Industrial Revolution. "There were too many social obstacles and too many technical difficulties for any general mechanization," says Bertrand Gille. "Nevertheless, the progress achieved was far from negligible and marks a considerable advance on the machinery of the ancient world."[152] It also promoted the evolution of the blast furnace, indispensable to supplying metal in a volume suitable to a mechanized industry. Such an industry also required skilled and knowledgeable workers; the invention of clockwork contributed to the skill and the invention of printing to the knowledge. As Terry Reynolds summarizes, "The roots from which the modern factory system emerged were quite deeply imbedded in the Middle Ages . . . there were no sharp breaks between the water-powered fulling and iron mills of the late Middle Ages and the textile mills of Strutt and Arkwright."[153]

In its organization of work, too, the high Middle Ages took a giant stride. The putting-out system—the "factory scattered through the town"—and its successor arrangements in Italy, England, and Germany clearly pointed the way to the future. In another dimension, so did the Venetian Arsenal and its pioneering "assembly line."

The Scientific Revolution of Galileo, Tycho, and Newton also profited from the intellectual and practical contributions of the Middle Ages, notably the invention of the convex lens. "In the Scientific Revolution of the sixteenth and seventeenth centuries," says Derek de Solla Price, "the dominant influences were the craft tradition and the printed book."[154] Thus technology served science, foreshadowing a future full partnership of the two. Nor should the role of medieval scientific thought be overlooked. The old picture of modern science springing directly from Aristotle and antiquity has lost validity: "Modern science . . . is rather the child of medieval science" (Richard Dales).[155]

Although the age succeeding that of Leonardo witnessed a relaxation of the pace of technical change, the perception of technology gained noticeably in stature, "capturing a place it had never before occupied" (Bertrand Gille).[156] For one thing, it had gained political importance, not only in the form of small arms and artillery but in many areas of mining, metallurgy, and craft production in which the new national governments interested themselves. For another, through the medium of printed books, technical information became diffused in the general body of knowledge.

Technology is rarely an unmixed blessing. The Middle Ages has been criticized by some modern historians for its depletion of the European forests. "Throughout the Middle Ages and Renaissance," says Carlo Cipolla, "the Europeans behaved toward the trees in an eminently parasitic and extremely wasteful way."[157] The judgment seems severe in light of medieval man's necessary dependence on trees for many purposes: building construction, tools, furniture, cooking and heating, the forge, the blast furnace, the bake oven, the pottery kiln, tile and brick making, glassmaking, distilling. In the absence of technological means for increasing yields, the only way to grow more crops to feed the increasing population was to cultivate more land. To provide arable land, and in the later Middle Ages meadowland, the forest had to give way. But as Roland Bechmann says in his study of medieval forest history, "It is during the Middle Ages that the idea of prospective planning and economical use of natural resources was gradually conceived by men who remained close to the land."[158] The Middle Ages was the first historical period to encounter the problem of limited natural resources, and it took the first small steps in dealing with it.

Besides royal and seigneurial protection of forests for hunting, many conservation initiatives are recorded, such as the enclosures established by the Cistercians in 1281 to protect seedlings and the mandating of tree planting by an Italian commune shortly after.[159] In the fourteenth century, French royal ordinances regulated cutting of timber; in the fifteenth, Venice's Council of Ten, concerned not only about deforestation but about the silting of the lagoons, strictly limited consumption of timber and prescribed planting of oak seedlings on common lands, with special sowings near the lagoons.[160]

That there were losses along with the gains in medieval technology is undeniable. Perhaps as much to be regretted as the destruction of forests was the other face of mechanization, the loss of handicraft skills, including those of the scribe and the artist who together created the illuminated manuscript.

"Technology," says Melvin Kranzberg, founder of the Society for the History of Technology, "is neither good nor bad; nor is it neutral."[161] It is what each age and each society make of it. The Middle Ages used it sometimes wisely, sometimes recklessly, often for dubious purposes, seldom with a thought for the future, and with only a dim awareness of the scientific and mathematical laws governing it. But operating on instinct, insight, trial and error, and perseverance, the craftsmen and craftswomen, the entrepreneurs, the working monks and the clerical intellectuals, and the artist-engineers all transformed the world, on balance very much to the world's advantage.

NOTES

1: NIMROD'S TOWER, NOAH'S ARK

1. Thomas Sprat, *History of the Royal Society*, London, 1667, p. 14.
2. William Wotton, *Reflections upon Ancient and Modern Learning*, London, 1894, pp. 1–3.
3. Edward Gibbon, *Decline and Fall of the Roman Empire*, New York, n.d. (Modern Library edition) (first pub. in 6 vols., 1776–1788), vol. II, p. 1443.
4. Jerome Cardan, *De subtilitate*, bk. 3, Nuremberg, 1550, cited in Joseph Needham, *Science and Civilization in China*, Cambridge, 1954–, vol. IV, pt. 2, *Mechanical Engineering*, p. 7.
5. Bern Dibner, ed., *The "New Discoveries": The Sciences, Inventions, and Discoveries of the Middle Ages and Renaissance as Represented in 24 Engravings Issued in the Early 1580s by Stradanus*, Norwalk, 1953.
6. Anne-Robert-Jacques Turgot, *On Progress, Sociology, and Economics*, ed. and trans. Ronald L. Meek, Cambridge, 1973, p. 55.
7. Fred C. Robinson, "Medieval, the Middle Ages," *Speculum* 59 (1984), pp. 745–56.
8. George Ovitt, Jr., *The Restoration of Perfection: Labor and Technology in Medieval Culture*, New Brunswick, N.J., 1987, p. 139.
9. Jean Gimpel, *The Medieval Machine*, New York, 1976.
10. Lewis Mumford, *Technics and Civilization*, London, 1934, p. 109.

11. Lynn White, Jr., "The Historical Roots of Our Ecological Crisis," *Science* 156 (1967), p. 1205.

12. Max Weber, *The Protestant Ethic and the Spirit of Capitalism*, trans. Talcott Parsons, New York, 1958, pp. 118–19.

13. Ernest Benz, "I fondamenti cristiani della tecnica occidentale," in *Tecnica e casistica*, ed. Enrico Castelli, Rome, 1964, pp. 241–63.

14. Ovitt, *Restoration of Perfection*, p. 40.

15. Ibid., p. 165.

16. Tertullian, *De anima*, in J. P. Migne, ed., *Patrologia latina*, Paris, 1844, vol. II, col. 700.

17. Augustine, *De civitate Dei*, ed. Bernard Dombast and Alphonse Kalb, *Corpus christianorum, series latina*, vol. 48, Turnhout, 1955, bk. 13, ch. 21, cited in Ovitt, *Restoration of Perfection*, p. 79.

18. Bede, *Opera exegetica: Libri quatuor in principium Genesis*, ed. C. W. Jones, *Corpus christianorum, series latina*, vol. 118A, Turnhout, 1967, p. 51, cited in Ovitt, *Restoration of Perfection*, p. 80.

19. Thomas Aquinas, *Summa theologiae*, ed. and trans. the Fathers of the English Dominican Province, London, 1964, vol. 13, p. 125 (part 1, question 96).

20. Palladius, *Historia Lausiaca*, in *Patrologiae cursus completus, series graeca*, ed. J. P. Migne, Paris, 1857–1904, vol. 34, 2.87–96, cols. 991–1278, cited in Ovitt, *Restoration of Perfection*, p. 94.

21. *Pachomiana latina*, ed. A. Boon, *Bibliothèque de la revue d'histoire ecclésiastique* 7 (Louvain, 1932), cited in Ovitt, *Restoration of Perfection*, p. 94.

22. St. Bernard, *Sermons*, sermon 26, cited in Ovitt, *Restoration of Perfection*, pp. 146–47.

23. Ovitt, *Restoration of Perfection*, pp. 144–45.

24. Walter Daniel, *Vita Aelredi*, ed. and trans. Maurice Powicke, Oxford, 1978, pp. 22–23.

25. Ovitt, *Restoration of Perfection*, p. 157.

26. Aristotle, *Politics*, trans. H. Rackham, London, 1932, pp. 638–39.

27. Aristotle, *Politics*, trans. Ernest Barker, Oxford, 1948, 1248b, 33ff.

28. Cicero, *De officiis*, London, 1913, p. 152, cited in Elspeth Whitney, *Paradise Restored: The Mechanical Arts from Antiquity Through the Thirteenth Century*, Philadelphia, 1990, p. 29.

29. Augustine, *City of God*, trans. Gerald G. Walsh, Demetrius B. Zema, Grace Monahan, and Daniel Honan, New York, 1958, bk. 22, ch. 24, p. 526.

30. *Letters of Cassiodorus*, trans. Thomas Hodgkin, London, 1886, p. 483, cited in Whitney, *Paradise Restored*, p. 67.

31. Isidore of Seville, *Etymologiae*, ed. W. M. Lindsay, Oxford, 1911, cited in Whitney, *Paradise Restored*, p. 66.

32. Whitney, *Paradise Restored*, pp. 72–73.

33. Honorius Augustodunensis, *De animae exsilio*, cited in Whitney, *Paradise Restored*, p. 69.

34. Hugh of St. Victor, *Hugonis de Sancto Victore Didascalicon de studio legendi: A Critical Text*, ed. Charles Henry Buttimer, Washington, D.C., 1939, pp. 39–40.

35. *The Didascalicon of Hugh of St. Victor: A Medieval Guide to the Arts*, trans. Jerome Taylor, New York, 1961, pp. 76–77.

36. Ibid., pp. 55–56.

37. Robert Kilwardby, *De ortu scientiarum*, ed. Albert G. Judy, Oxford, 1976, pp. 139–40.

38. Roger Bacon, *Opus maius*, trans. Robert Belle Burke, New York, 1962, p. 36.

39. Richard Lefebvre des Noëttes, *L'Attelage et le cheval de selle à travers les ages*, Paris, 1931.

40. Lynn White, Jr., *Medieval Technology and Social Change*, London, 1962.

41. Peter F. Drucker, "The First Technological Revolution and Its Lessons," *Technology and Culture* 7 (1966), pp. 143–51; John H. Meursinge, "Overlapping Histories of Technology," *Technology and Culture* 8 (1967), pp. 517–18.

42. Needham, *Science and Civilization*, vol. I, *Introductory Orientations*, p. 244.

43. Joseph Glanvill, *Scepsis scientifica, or Confest Ignorance, the Way to Science*, London, 1665, p. 140.

2: THE TRIUMPHS AND FAILURES OF ANCIENT TECHNOLOGY

1. Bertrand Gille, ed., *The History of Techniques*, vol. I, *Techniques and Civilizations*, New York, 1986 (originally pub. in 1978 as *Histoire des techniques*), p. 147.

2. Homer, *Odyssey*, trans. Robert Fitzgerald, New York, 1961, bk. 2, lines 94–95.

3. Jean Deshayes, "Greek Technology," in Maurice Daumas, ed., *A History of Technology and Invention: Progress Through the Ages*, trans. Eileen B. Hennessy, New York, 1970 (henceforth referred to as Daumas), vol. I, pp. 187, 196; R. J. Forbes and E. J. Dijksterhuis, *A History of Science and Technology*, vol. I, *Ancient Times to the Seventeenth Century*, Harmondsworth, 1963 (henceforth

referred to as Forbes and Dijksterhuis), pp. 67–68; L. Sprague de Camp, *The Ancient Engineers*, New York, 1963, pp. 70–71.

4. De Camp, *Ancient Engineers*, p. 141.

5. T. K. Derry and Trevor I. Williams, *A Short History of Technology from the Earliest Times to A.D. 1900*, Oxford, 1960 (henceforth referred to as Derry and Williams), pp. 120–21; Georges Contenau, "Mesopotamia and the Neighboring Countries," in Daumas, I, p. 136.

6. Forbes and Dijksterhuis, p. 72.

7. De Camp, *Ancient Engineers*, pp. 39–40, 43, 92, 93; B. Gille, *History of Techniques*, I, pp. 257–58.

8. B. Gille, *History of Techniques*, I, p. 264.

9. T. C. Lethbridge, "Shipbuilding," in Charles Singer, E. J. Holmyard, A. R. Hall, and Trevor I. Williams, eds., *A History of Technology*, Oxford, 1954–1959, 1978 (henceforth referred to as Singer), vol. II, *The Mediterranean Civilizations and the Middle Ages, 700 B.C. to A.D. 1500*, p. 564.

10. Homer, *Odyssey*, bk. 5, lines 244–51.

11. Homer, *Odyssey*, bk. 15, line 403, cited in Sabatino Moscati, *The World of the Phoenicians*, New York, 1965, p. 87.

12. M. I. Finley, "Technical Innovation and Economic Progress in the Ancient World," *Economic History Review*, 2nd ser., 18 (1965), p. 32.

13. Deshayes, "Greek Technology," in Daumas, I, p. 191; J. G. Landels, *Engineering in the Ancient World*, Berkeley, Calif., 1978, p. 59; Donald Hill, *A History of Engineering in Classical and Medieval Times*, La Salle, Ill., 1984, pp. 132–33, citing A. G. Drachmann, *The Mechanical Technology of Greek and Roman Antiquity*, Madison, Wis., 1963, p. 154.

14. Landels, *Engineering*, p. 76; Abbott Payson Usher, *A History of Mechanical Inventions*, Boston, 1959 (first pub. in 1929), pp. 134–36.

15. A. J. Turner, *Astrolabes; Astrolabe-Related Instruments*, Rockford, Ill., 1985; J. D. North, "The Astrolabe," *Scientific American* 230 (1974), pp. 96–106. A far more complex device dating from the first century B.C. was found off the island of Antikythera early in the twentieth century and identified by Derek de Solla Price as an elaborate astronomical calendar. Price pointed out that the device exploded the myth that the Greeks were weak in technology: "The technology was there . . . It has just not survived like

the great marble buildings and the constantly recopied literary works." (Derek de Solla Price, *Science Since Babylon*, New Haven, 1976, p. 48.)

16. Paul-Marie Duval, "The Roman Contribution to Technology," in Daumas, I, pp. 245–46; E. M. Jope, "Agricultural Implements," in Singer, II, p. 86; R. Z. Patterson, "Spinning and Weaving," in Singer, II, p. 193.

17. De Camp, *Ancient Engineers*, p. 227; Landels, *Engineering*, p. 15; R. J. Forbes, "Food and Drink," in Singer, II, p. 117; Deshayes, "Greek Technology," in Daumas, I, p. 211; Duval, "Roman Contribution," in Daumas, I, pp. 245–46.

18. Derry and Williams, pp. 123–24; Deshayes, "Greek Technology," in Daumas, I, pp. 198–99; C. N. Bromehead, "Mining and Quarrying to the Seventeenth Century," in Singer, II, pp. 3–7.

19. Duval, "Roman Contribution," in Daumas, I, pp. 242–43.

20. J. P. Wild, *Textile Manufacture in the Northern Roman Provinces*, Cambridge, 1970, pp. 35–36, 61–72.

21. Duval, "Roman Contribution," in Daumas, I, p. 232.

22. Ibid., pp. 219, 226; R. J. Forbes, "Hydraulic Engineering and Sanitation," in Singer, II, pp. 670–71.

23. Joseph Gies, *Bridges and Men*, New York, 1963, pp. 8–11.

24. Duval, "Roman Contribution," in Daumas, I, pp. 247, 223, 228–29.

25. Ibid., p. 224; R. G. Goodchild, "Roads and Land Travel, with a Section on Harbours, Docks, and Lighthouses," in Singer, II, pp. 500–514.

26. Landels, *Engineering*, pp. 136–42; Lionel Casson, "Odysseus' Boat (*Od.* V, 244–57)," *American Journal of Philology* 85 (1964), pp. 86–90; Duval, "Roman Contribution," in Daumas, I, pp. 238–39; Derry and Williams, p. 197.

27. Landels, *Engineering*, pp. 157–58; Lynn White, Jr., *Medieval Religion and Technology: Collected Essays*, Berkeley, Calif., 1978, pp. 255–60. Lionel Casson lists five types of ancient fore-and-aft sail: the triangular lateen of the Mediterranean; the quadrilateral ("Arab") lateen; the spritsail; the gaff-headed sail; the lugsail. All were apparently limited to small craft. (*Ships and Seamanship in the Ancient World*, Princeton, N.J., 1971, p. 243.)

28. Landels, *Engineering*, pp. 107–9.

29. Derry and Williams, p. 58.

30. Duval, "Roman Contribution," in Daumas, I, p. 245.

31. Finley, "Technical Innovation," p. 30; Forbes and Dijksterhuis, p. 81.
32. Duval, "Roman Contribution," in Daumas, I, pp. 251–52; B. Gille, *History of Techniques*, I, pp. 427–28.
33. Duval, "Roman Contribution," in Daumas, I, p. 254.
34. Derry and Williams, p. 61; Kenneth Kilby, *The Cooper and His Trade*, London, 1971, p. 95.
35. Needham, *Science and Civilization*, vol. IV, pt. 2, pp. 318–19.
36. Lefebvre des Noëttes, *L'Attelage et le cheval de selle*, p. 5.
37. Terry S. Reynolds, *Stronger Than a Hundred Men: A History of the Vertical Water Wheel*, Baltimore, 1983, p. 14; Derry and Williams, pp. 250–52; R. J. Forbes, "Power," in Singer, II, pp. 590–600; André Haudricourt and Maurice Daumas, "The First Stages in the Utilization of Natural Power," in Daumas, I, pp. 108–9.
38. Derry and Williams, p. 32.
39. Vitruvius, *De architectura*, 10.1.6, cited in T. Reynolds, *Stronger Than a Hundred Men*, p. 30.
40. T. Reynolds, *Stronger Than a Hundred Men*, p. 11.
41. Forbes and Dijksterhuis, p. 76; T. Reynolds, *Stronger Than a Hundred Men*, p. 31.
42. White, *Medieval Religion and Technology*, p. 225.
43. Seneca, *On Mercy*, VII, 25, 2–4, cited in De Camp, *Ancient Engineers*, p. 254.
44. William D. Phillips, Jr., *Slavery from Roman Times to the Early Transatlantic Trade*, Minneapolis, 1985, pp. 22–23.
45. Finley, "Technical Innovation," p. 37.
46. Ibid., p. 29.

3: THE NOT SO DARK AGES: A.D. 500–900

1. Gibbon, *Decline and Fall of the Roman Empire*, vol. II, p. 9.
2. Henri Pirenne, *Mohammed and Charlemagne*, trans. Bernard Miall, New York, 1964 (first pub. in 1939), p. 140.
3. White, *Medieval Religion and Technology*, p. 12.
4. Derry and Williams, p. 91.
5. R. J. Forbes, "Metallurgy," in Singer, II, p. 62.
6. White, "Technology and Invention in the Middle Ages," *Speculum* 15 (1940), p. 151.
7. Carlo Cipolla, *Before the Industrial Revolution*, New York, 1980, p. 169.

8. Georges Duby, *The Early Growth of the European Economy: Warriors and Peasants from the Seventh to the Twelfth Century*, trans. Howard B. Clarke, Ithaca, N.Y., 1974, pp. 13, 29.

9. Cipolla, *Before the Industrial Revolution*, p. 113.

10. Duby, *Early Growth*, p. 3.

11. Richard Unger, *The Ship in the Medieval Economy, 600–1600*, London, 1980, p. 97.

12. Duby, *Early Growth*, pp. 8–9.

13. Robert S. Lopez, *The Commercial Revolution of the Middle Ages, 950–1350*, Cambridge, 1976, pp. 16–17.

14. Richard Hodges and David Whitehouse, *Mohammed, Charlemagne, and the Origins of Europe: Archaeology and the Pirenne Thesis*, Ithaca, N.Y., 1983, pp. 92–101.

15. Gustave Milne and Damian Goodburn, "The Early Medieval Port of London, A.D. 700–1200," *Antiquity* 64 (1990), pp. 629–36.

16. David Hill, David Barrett, Keith Maude, Julia Warburton, and Margaret Worthington, "Quentovic Defined," *Antiquity* 64 (1990), pp. 51–58.

17. Frances and Joseph Gies, *Marriage and the Family in the Middle Ages*, New York, 1987, pp. 45–46.

18. White, *Medieval Technology and Social Change*, p. 56.

19. Ibid., p. 61.

20. Albert C. Leighton, *Transport and Communication in Early Medieval Europe, A.D. 500–1100*, Newton Abbot, 1971, pp. 108–12.

21. White, *Medieval Technology and Social Change*, p. 63.

22. Frances and Joseph Gies, *Life in a Medieval Village*, New York, 1990, p. 59.

23. *The Rule of St. Benedict*, cited in Ovitt, *Restoration of Perfection*, p. 103.

24. Ibid., p. 104.

25. Gregory of Tours, *The History of the Franks*, trans. Lewis Thorpe, Harmondsworth, 1974, ch. III, sect. 19, p. 182.

26. Gregory of Tours, *Lives of the Fathers (Liber vitae patrum)*, cited in Philip A. Rahtz and Donald Bullough, "The Parts of an Anglo-Saxon Mill," *Anglo-Saxon England* 6 (1977), p. 20.

27. Katherine Fischer Drew, ed. and trans., *The Laws of the Salian Franks*, Philadelphia, 1991, p. 85.

28. Katherine Fischer Drew, ed. and trans., *The Lombard Laws*, Philadelphia, 1973, pp. 76–77.

29. T. Reynolds, *Stronger Than a Hundred Men*, p. 49; Forbes, "Power," in Singer, II, pp. 607–13.

30. Georges Duby, *Rural Economy and Country Life in the Medieval West*, trans. Cynthia Postan, Columbia, S.C., 1968, pp. 16–17.

31. Ibid., p. 17.

32. Richard Holt, *The Mills of Medieval England*, London, 1988, p. 123.

33. Charlemagne's *Admonitio generalis*, cited in Robert Latouche, *The Birth of Western Economy: Economic Aspects of the Dark Ages*, trans. E. M. Wilkinson, New York, 1961, p. 176.

34. Gregory of Tours, *History*, ch. IX, sect. 38, p. 525.

35. David Herlihy, *Opera Muliebria: Women and Work in Medieval Europe*, Philadelphia, 1990, pp. 42–43.

36. David Herlihy, ed., *Medieval Culture and Society*, New York, 1968, p. 48.

37. Herlihy, *Opera Muliebria*, pp. 75–81; Walter Endrei, *L'Evolution des techniques du filage et du tissage du Moyen Age à la révolution industrielle*, trans. from the Hungarian by Joseph Tackacs, Paris, 1968, pp. 30–31.

38. Needham, *Science and Civilization*, vol. I, pp. 185–86.

39. Einhard and Notker the Stammerer, *Two Lives of Charlemagne*, trans. Lewis Thorpe, Harmondsworth, 1979, p. 147.

40. Endrei, *Evolution des techniques*, p. 12.

41. Wild, *Textile Manufacture*, pp. 35–36.

42. Herlihy, *Opera Muliebria*, p. 80.

43. Wild, *Textile Manufacture*, pp. 61–68; Endrei, *Evolution des techniques*, pp. 24–25; Marta Hoffmann, *The Warp-weighted Loom: Studies in the History and Technique of an Ancient Implement*, Oslo, 1964.

44. Wild, *Textile Manufacture*, pp. 28–29; Patterson, "Spinning and Weaving," in Singer, II, p. 196.

45. White, *Medieval Technology and Social Change*, pp. 1–38.

46. Bernard S. Bachrach, "Charles Martel, Mounted Shock Combat, the Stirrup, and Feudal Origins," *Studies in Medieval and Renaissance History* 7 (1970), pp. 47–76; Bernard S. Bachrach, *Merovingian Military Organization, 481–751*, Minneapolis, 1972, pp. 113–28; P. H. Sawyer and R. H. Hilton, "Technical Determinism: The Stirrup and the Plough," review of White, *Medieval Technology and Social Change*, in *Past and Present* 24 (1963), pp. 90–95; Frances Gies, *The Knight in History*, New York, 1984, pp. 9–12.

47. White, *Medieval Technology and Social Change*, pp. 14–24; A. D. H. Bivar, "Cavalry Equipment and Tactics on the Euphrates Frontier," *Dumbarton Oaks Papers* 26 (1979), pp. 273–91; A. D. H. Bivar, "The Stirrup and Its Origins," *Oriental Art*, n.s. 1 (1955), pp. 61–65.

48. Bachrach, "Charles Martel," pp. 59–60.

49. David C. Douglas, *William the Conqueror*, Berkeley, Calif., 1967, p. 202.

50. Leighton, *Transport and Communication*, pp. 106–8.

51. Translation adapted from Einhard and Notker, *Two Lives of Charlemagne*, pp. 163–64, and Bachrach, "Charles Martel," p. 61.

52. Philippe Contamine, *War in the Middle Ages*, trans. Michael Jones, London, 1984, p. 58.

53. Jean de Colmieu, *Vie de Jean de Warneton*, cited in Sidney Toy, *The Castles of Great Britain*, London, 1953, pp. 44–45.

54. Lambert of Ardres, *Historia comitum Ghisnensium*, in *Monumenta Germaniae historica scriptores*, ed. G. H. Pertz et al., Hanover, 1826–1913, vol. XXIV, ch. 127, p. 624.

55. Needham, *Science and Civilization*, vol. V, *Chemistry and Chemical Technology*, pt. 7, *Military Technology: The Gunpowder Epic*, pp. 73–89; Alex Roland, "Secrecy, Technology, and War: Greek Fire and the Defense of Byzantium, 679–1204," *Technology and Culture* 33 (1992), pp. 655–79; *The Chronicle of Theophanes*, trans. Harry Turtledove, Philadelphia, 1982, pp. 88–89.

56. Roland Bechmann, *Trees and Man: The Forest in the Middle Ages*, trans. Katharyn Dunham, New York, 1990, p. 152; John Hooper Harvey, *Mediaeval Craftsmen*, London, 1975.

57. Michael Swanton, ed., *Anglo-Saxon Prose*, London, 1975, p. 113.

58. Bertrand Gille, "The Transformation of Raw Materials," in Daumas, I, pp. 493–94; David W. Crossley, "Medieval Iron Smelting," in Crossley, ed., *Medieval Industry* (Research Report 40, Council for British Archaeology), London, 1981, pp. 33–34; Jane Geddes, "Iron," in John W. Blair and Nigel Ramsay, eds., *English Medieval Industries*, London, 1990, pp. 168–73.

59. W. K. V. Gale, *Iron and Steel*, London, 1969, p. 12; Arnold Pacey, *The Maze of Ingenuity: Ideas and Idealism in the Development of Technology*, Cambridge, Mass., 1992, p. 12; B. Gille, "Transformation of Raw Materials," in Daumas, I, pp. 493–95.

60. Geddes, "Iron," p. 173.

61. Leslie Aitchison, *A History of Metals*, London, 1960, vol. I, pp. 248–49.

62. Ibid., p. 142.

63. Ibid., pp. 253–54.

64. Bertrand Gille, "The Problems of Power and Mechanization," in Daumas, I, p. 448.

65. White, *Medieval Technology and Social Change*, p. 110.

66. Bertrand Gille, "La Naissance du système bielle-manivelle," *Techniques et civilisations* 2 (1952), pp. 42–46; Needham, *Science and Civilization*, vol. IV, pt. 2, pp. 14–16.

67. Jean Theodoridès, "The Byzantine Empire (Sixth to Fifteenth Centuries)," in Daumas, I, p. 373.

68. Robert L. Reynolds, *Europe Emerges: Transition Toward an Industrial World-wide Society, 600–1750*, Madison, Wis., 1967, p. 33.

69. Gregory of Tours, *History*, ch. II, sect. 14, p. 130; Pierre Lavedan, *French Architecture*, Harmondsworth, 1956, p. 83.

70. Bertrand Gille, "The Organization of Space," in Daumas, I, pp. 529–30; Lavedan, *French Architecture*, p. 84.

71. Bede, *A History of the English Church and People*, trans. Leo Sherley-Price, Harmondsworth, 1986, p. 315.

72. B. Gille, "Organization of Space," in Daumas, I, p. 546.

73. Leicester Bodine Holland, "Traffic Ways About France in the Dark Ages (500–1150)," Ph.D. thesis, University of Pennsylvania, Allentown, 1919, p. 6.

74. Bertrand Gille, "The Problem of Transportation," in Daumas, I, p. 436.

75. Marjorie Nice Boyer, *Medieval French Bridges: A History*, Cambridge, Mass., 1976, p. 17.

76. Holland, "Traffic Ways," pp. 63, 71.

77. Boyer, *Medieval French Bridges*, p. 160.

78. C. T. Flower, *Public Works in Medieval Law*, London, 1915–1923, cited in Boyer, *Medieval French Bridges*, p. 160.

79. Forbes, "Power," in Singer, II, pp. 607–8.

80. Unger, *The Ship in the Medieval Economy*, p. 47; John H. Pryor, *Geography, Technology, and War: Studies in the Maritime History of the Mediterranean, 649–1571*, Cambridge, 1988, p. 27. That Muslim and Christian ships resembled each other is indicated by the popular *ruse de guerre* of concealing a ship's identity until close to an enemy or pirate prey.

81. Unger, *The Ship in the Medieval Economy*, pp. 47, 49.

82. Pryor, *Geography, Technology, and War*, p. 27.

83. Leighton, *Transport and Communication*, p. 143.

84. Unger, *The Ship in the Medieval Economy*, p. 63; B. Gille, "Problem of Transportation," in Daumas, I, pp. 437–38; T. C. Lethbridge, "Shipbuilding," in Singer, II, p. 579.

85. Gwyn Jones, *A History of the Vikings*, New York, 1968, pp. 186–89; Lethbridge, "Shipbuilding," in Singer, II, p. 580.

86. Unger, *The Ship in the Medieval Economy*, p. 88.

87. Jones, *History of the Vikings*, p. 187.

88. Ibn Fadlan, cited in Hodges and Whitehouse, *Mohammed, Charlemagne, and the Origins of Europe*, p. 123, and Jones, *History of the Vikings*, pp. 164–65.

89. Richard Unger, "Warships and Cargo Ships in Medieval Europe," *Technology and Culture* 22 (1981), p. 242.

90. Ibid., p. 240.

91. Leighton, *Transportation and Communication*, pp. 15–16.

92. R. J. Forbes, *Man the Maker: A History of Technology and Engineering*, London, 1958, p. 105.

93. Boethius, *Consolation of Philosophy*, trans. Richard Green, bk. 1, poem 2, pp. 5–6, cited in John F. Benton, "Ideas of Order: Music, Mathematics, and Medieval Architecture," Caltech Lecture Series, p. 6.

94. Ibid., p. 18.

95. Charles Homer Haskins, *The Renaissance of the Twelfth Century*, New York, 1963 (first pub. in 1927), p. 33.

96. Pierre Riché, *Daily Life in the World of Charlemagne*, trans. Jo Ann McNamara, Philadelphia, 1980, p. 208.

97. Lynn White, Jr., "Dynamo and Virgin Reconsidered," *American Scholar* 27 (1958), p. 189.

98. Riché, *Daily Life*, p. 145.

99. James Bryce, *The Holy Roman Empire*, New York, 1905, p. 80.

100. P. Boissonade, *Life and Work in Medieval Europe: The Evolution of Medieval Economy from the Fifth to the Fifteenth Century*, trans. Eileen Power, New York, 1964 (first pub. in 1927), p. 95.

101. R. Reynolds, *Europe Emerges*, p. 156.

102. Lynn White, Jr., "Conclusion: The Temple of Jupiter Revisited," in White, ed., *The Transformation of the Roman World: Gibbon's Problem After Two Centuries*, Berkeley, Calif., 1966, p. 304.

4: THE ASIAN CONNECTION

1. Needham, *Science and Civilization*, vol. I, pp. 161–62.

2. Ibid., p. 238.

3. Ibid., p. 168.
4. Wild, *Textile Manufacture*, pp. 26–27.
5. Needham, *Science and Civilization*, vol. I, p. 187.
6. Ibid., p. 236.
7. Ibid., pp. 234–36.
8. Ibid., IV, pt. 2, p. 236; A. C. Crombie, *Medieval and Early Modern Science*, vol. I, *Science in the Middle Ages: V–XIII Centuries*, New York, 1959 (first pub. in 1952), p. 25.
9. Joseph Needham, "Poverties and Triumphs of the Chinese Scientific Tradition," in A. C. Crombie, ed., *Scientific Change*, New York, 1963, pp. 125, 131–32; Needham, *Science and Civilization*, vol. IV, pt. 2, p. 17.
10. Needham, "Poverties and Triumphs," pp. 126–31.
11. Needham, *Science and Civilization*, vol. IV, pt. 2, p. 19.
12. Ibid., vol. IV, pt. 1, *Physics*, pp. 239–40.
13. Ibid., vol. IV, pt. 2, p. 20.
14. Ibid., vol. I, pp. 244–48.
15. Ibid., p. 241.
16. Joseph Needham, "Chinese Priorities in Cast Iron Metallurgy," *Technology and Culture* 5 (1964), pp. 402–3.
17. Needham, *Science and Civilization*, vol. IV, pt. 2, pp. 370–71.
18. Ibid., pp. 192–94.
19. Ibid., pp. 364–65.
20. Ibid., pp. 392–94.
21. Ibid., p. 416. The idea of the paddle-wheel boat was present in the Roman treatise *De rebus bellicis*, now believed to date to about A.D. 370, but the boat was evidently never actually built. E. A. Thompson, ed., *A Roman Reformer and Inventor, Being a Text of the Treatise "De rebus bellicis*," Oxford, 1950; also Needham, *Science and Civilization*, vol. IV, pt. 2, pp. 413, 434.
22. Needham, *Science and Civilization*, vol. IV, pt. 2, pp. 447–64.
23. Ibid., p. 436.
24. Ibid., pp. 261–71.
25. Ibid., vol. I, pp. 230–31.
26. Ibid., p. 134.
27. Ibid., p. 129.
28. Ibid., vol. IV, pt. 2, pp. 282–84.
29. Ibid., pp. 290–95.
30. Ibid., pt. 1, p. 269.
31. Ibid., pp. 259–60, 281–82.

32. Ibid., pp. 249–51.

33. Ibid., p. 279.

34. Ibid., pt. 2, p. 233.

35. Ibid., vol. V, pt. 7, p. 1.

36. Ibid., pp. 14–15.

37. Ibid., pt. 1, *Paper and Printing* (Ysien Tsuen-Hsuin), p. 296.

38. Ibid., pp. 71–72, 73–76.

39. Ibid., pp. 109, 123.

40. Ibid., pp. 96–102. Marco Polo visited the mint in the Mongol capital of Kanbalu (modern Peking) and watched the paper currency being printed and issued. (Henry H. Hart, *Marco Polo, Venetian Adventurer*, Norman, Okla., 1967, pp. 118–19.) William of Rubruck described it as "the length and breadth of a palm [of the hand], stamped with lines similar to those of the seal of Mangu Khan." (Manuel Komroff, ed., *Contemporaries of Marco Polo*, New York, 1989 [first pub. in 1928], p. 152.)

41. Needham, *Science and Civilization*, vol. V, pt. 1, p. 365 (poem translated by Howard Winger).

42. Ibid., vol. I, p. 236; vol. V, pt. 1, pp. 297–98.

43. Derry and Williams, pp. 232–33.

44. Arnold Pacey, *Technology in World Civilization*, Cambridge, Mass., 1991, pp. 42–43.

45. Needham, *Science and Civilization*, vol. V, pt. 1, pp. 306–7.

46. Ibid., pp. 201–2.

47. Ibid., pp. 206–7.

48. Ibid., p. 304.

49. Joel Mokyr, *The Lever of Riches: Technological Creativity and Economic Progress*, New York, 1990, p. 39.

50. Andrew H. Watson, "The Arab Agricultural Revolution and Its Diffusion, 700–1100," *Journal of Economic History* 34 (1974), pp. 21–22.

51. Philip Hitti, *History of the Arabs from the Earliest Times to the Present*, London, 1964, pp. 309–15.

52. Bernard Lewis, *The Arabs in History*, New York, 1960 (first pub. in 1950), p. 91.

53. Charles Homer Haskins, *The Normans in European History*, New York, 1966 (first pub. in 1915), p. 228.

54. Aziz S. Atiya, *Crusade, Commerce, and Culture*, New York 1966 (first pub. in 1962), pp. 236, 238.

55. Watson, "Arab Agricultural Revolution," pp. 8–35.

56. Maureen Fennell Mazzaoui, *The Italian Cotton Industry in the Later Middle Ages (1100–1600)*, Cambridge, 1981, pp. 21–24; Atiya, *Crusade, Commerce and Culture*, p. 239.

5: THE TECHNOLOGY OF THE COMMERCIAL REVOLUTION: 900–1200

1. Pacey, *Technology in World Civilization*, p. 20.
2. Ibid., p. 41.
3. Jones, *History of the Vikings*, pp. 295–300.
4. Lethbridge, "Shipbuilding," in Singer, II, p. 581: "Were it not for the Norse custom of handing down stories of the lives of some of their prominent men, and their habit of burying their chieftains in ships, we should probably know no more about their widespread voyages than we do of those of their Irish predecessors. It seems probable that the Irish reached at least as far as Iceland, Greenland, Newfoundland, and the Azores ... in large, skin-covered boats holding twenty to thirty men apiece. St. Brendan (484–577) was the most famous of these Irish explorers."
5. R. Reynolds, *Europe Emerges*, pp. 185–86.
6. Herbert Heaton, *Economic History of Europe*, New York, 1936, p. 151.
7. Howard Saalman, *Medieval Cities*, New York, 1968, p. 114.
8. E. Barthélemy, *Notice historique sur les communes du canton de Ville-sur-Tourbe*, Paris, 1865, cited in Bechmann, *Trees and Man*, p. 104.
9. Robert S. Lopez and Irving W. Raymond, eds., *Medieval Trade in the Mediterranean World*, New York, 1955, pp. 162–84; Joseph and Frances Gies, *Life in a Medieval City*, New York, 1969, pp. 211–23; Joseph and Frances Gies, *Merchants and Moneymen: The Commercial Revolution, 1000–1500*, New York, 1971, pp. 75–82.
10. Gimpel, *The Medieval Machine*, p. 57.
11. Carl Stephenson, "In Praise of Medieval Tinkers," *Journal of Economic History* 8 (1948), p. 29.
12. F. and J. Gies, *Life in a Medieval Village*, pp. 14–18.
13. Ibid., pp. 129–35.
14. Mary Gies Hatch, "De gulzige Waterwolf: Medieval Dikes in Friesland," unpublished paper; Forbes and Dijksterhuis, pp. 141–42.
15. Duby, *Early Growth*, p. 187.
16. Forbes, "Power," in Singer, II, p. 609.
17. Duby, *Early Growth*, p. 187.

18. T. Reynolds, *Stronger Than a Hundred Men*, p. 119; Derry and Williams, p. 253.

19. Bradford B. Blaine, "The Enigmatic Water-Mill," in Bert S. Hall and Delno C. West, eds., *On Pre-modern Technology and Science: A Volume of Studies in Honor of Lynn White, Jr.*, Malibu, Calif., 1976, pp. 167–69.

20. Forbes, "Power," in Singer, II, p. 610; B. Gille, "Problems of Power and Mechanization," in Daumas, I, p. 455.

21. White, *Medieval Religion and Technology*, p. 245.

22. Forbes, "Power," in Singer, II, p. 610.

23. E. M. Carus-Wilson, "An Industrial Revolution of the Thirteenth Century," *Economic History Review* 12 (1941), pp. 39–60; Forbes, "Power," in Singer, II, p. 611.

24. T. Reynolds, *Stronger Than a Hundred Men*, p. 83.

25. B. Gille, "Problems of Power and Mechanization," in Daumas, I, p. 455; T. Reynolds, *Stronger Than a Hundred Men*, pp. 79–81.

26. T. Reynolds, *Stronger Than a Hundred Men*, p. 106; Forbes, "Power," in Singer, II, p. 590.

27. Holt, *Mills of Medieval England*, pp. 37–69. Lynn White pictured the water mill as part of the Middle Ages' "humanitarian technology," a "labor-saving power-machine" produced by "an instinctive repugnance toward subjecting any man to a monotonous drudgery which seems less than human." ("Technology and Invention in the Middle Ages," p. 156.) Pierre Dockès, in contrast, regarded it as purely an instrument of exploitation, "above all a way of redistributing income, increasing the surplus that accrued to the masters ... It was practically never in the interest of the peasant to use it." Technical progress in general, in his view, was and remains "a by-product of social struggles" and an incidental feature of man's exploitation of man. (*Medieval Slavery and Liberation*, trans. Arthur Goldhammer, Chicago 1982, pp. 178–82.) The truth seems to lie somewhere between the two views.

28. Hill, *History of Engineering*, p. 58.

29. Ibid., p. 60; T. Reynolds, *Stronger Than a Hundred Men*, p. 65.

30. T. Reynolds, *Stronger Than a Hundred Men*, p. 63.

31. Holt, *Mills of Medieval England*, p. 133; T. Reynolds, *Stronger Than a Hundred Men*, p. 67.

32. Herlihy, *Opera Muliebria*, pp. 91–94.

33. Urban Tigner Holmes, Jr., *Daily Living in the Twelfth Century*,

Based on the Observations of Alexander Neckam in London and Paris, Madison, Wis., 1966 (first pub. in 1952), pp. 146–48.

34. Thomas Wright, ed., *A Volume of Vocabularies*, London, 1857, p. 106.

35. Alexander Neckam, *De naturis rerum*, trans. Thomas Wright, London, 1863, p. 281.

36. Herlihy, *Opera Muliebria*, p. 95.

37. Lopez, *Commercial Revolution*, p. 160.

38. Endrei, *Evolution des techniques*, pp. 43–44.

39. Robert S. Lopez, "Still Another Renaissance?" *American Historical Review* 57 (1951), p. 12.

40. John H. Munro, "The Medieval Scarlet and the Economics of Sartorial Splendour," in N. B. Harte and K. G. Ponting, *Cloth and Clothing in Medieval Europe, Essays in Memory of Professor E. M. Carus-Wilson*, London, 1983, p. 13.

41. Mazzaoui, *Italian Cotton Industry*, pp. 74–77.

42. Charles Singer, "Epilogue: East and West in Retrospect," in Singer, II, p. 762.

43. Endrei, *Evolution des techniques*, p. 47; Steven Runciman, *Byzantine Civilization*, New York, 1956, p. 135.

44. R. Reynolds, *Europe Emerges*, p. 226.

45. Chrétien de Troyes, *Le Conte del Graal*, verses 5765ff., cited in Holmes, *Daily Living*, pp. 133–34.

46. John W. Waterer, "Leather," in Singer, II, pp. 144–58.

47. Lopez, *Commercial Revolution*, p. 126.

48. Harvey, *Mediaeval Craftsmen*, p. 12.

49. F. Sherwood Taylor, "Pre-scientific Industrial Chemistry," in Singer, II, p. 356.

50. Judith M. Bennett, "The Village Ale-Wife: Women and Brewing in Fourteenth Century England," in Barbara A. Hanawalt, ed., *Women and Work in Preindustrial Europe*, Bloomington, Ind., 1986, pp. 20–36; Forbes, "Food and Drink," in Singer, II, p. 141.

51. Eileen Power, *Medieval Women*, ed. by M. M. Postan, Cambridge, 1975, p. 59.

52. Duby, *Early Growth*, pp. 194–95.

53. R. H. G. Thomson, "The Medieval Artisan," in Singer, II, p. 394.

54. Geddes, "Iron," pp. 175–77.

55. John G. Hawthorne and Cyril Stanley Smith, *On Divers Arts: The Treatise of Theophilus*, Chicago, 1963; Bechmann, *Trees and Man*, p. 172.

56. Hawthorne and Smith, *On Divers Arts*, p. 97.

57. Nadine George, "Albertus Magnus and Chemical Technology in a Time of Transition," in James A. Weisheipl, ed., *Albertus Magnus and the Sciences: Commemorative Essays*, Toronto, 1980, p. 240.

58. Lopez, *Commercial Revolution*, p. 143.

59. Vaclav Husa, Josef Petrau, and Alena Surbota, *Traditional Crafts and Skills: Life and Work in Medieval and Renaissance Times*, London, 1967, p. 152.

60. Pacey, *Maze of Ingenuity*, pp. 18, 25.

61. Nikolaus Pevsner, *An Outline of European Architecture*, Harmondsworth, 1954, p. 43.

62. B. Gille, "Organization of Space," in Daumas, I, p. 530.

63. Pevsner, *Outline of European Architecture*, p. 48.

64. Henri Daniel-Rops, *Cathedral and Crusade*, London, 1956, p. 96.

65. Erwin Panofsky, ed. and trans., *Abbot Suger on the Abbey Church of St. Denis and Its Art Treasures*, Princeton, N.J., 1946.

66. William of Malmesbury, *Chronicle of the Kings of England*, ed. and trans. J. A. Giles, London, 1889, p. 138.

67. J. and F. Gies, *Life in a Medieval City*, pp. 149–51; J. R. Hunter, "The Medieval Glass Industry," in Crossley, *Medieval Industry*, pp. 144–45.

68. Hawthorne and Smith, *On Divers Arts*, p. 57; Derry and Williams, pp. 94–95.

69. Hunter, "Medieval Glass Industry," p. 147.

70. B. Gille, "Organization of Space," in Daumas, I, p. 535.

71. Panofsky, *Abbot Suger*, excerpted in Bryce Lyon, ed., *The High Middle Ages, 1000–1300*, New York, 1964, p. 219.

72. Gervase of Canterbury, "Tract on the Burning and Repair of the Church of Canterbury," in R. Willis, *The Architectural History of Canterbury Cathedral*, London, 1945, excerpted in Lyon, ed., *High Middle Ages*, pp. 220–32.

73. "Architecture and Printing; the Bible of stone and the Bible of paper"—Victor Hugo, *Notre-Dame de Paris*, trans. Jessie Haynes, New York, 1955 (first pub. in 1831), p. 118. The metaphor has been used by many writers, for example, Daniel-Rops, *Cathedral and Crusade*, p. 101: "The Bible of colour went hand in hand with, and sometimes preceded, the Bible of stone."

74. Joseph and Frances Gies, *Life in a Medieval Castle*, New York, 1974, pp. 21–24; B. Gille, "Organization of Space," in Daumas, I, pp. 539–40.

75. Contamine, *War in the Middle Ages*, p. 109.
76. Atiya, *Crusade, Commerce, and Culture*, pp. 125–26.
77. Ibid., pp. 66–67; F. Gies, *Knight in History*, pp. 116–18; Robin R. Fedden and John Thomson, *Crusader Castles*, London, 1957; T. S. R. Boase, "Military Architecture in the Crusader States in Palestine and Syria," in Kenneth Setton, ed., *A History of the Crusades*, Madison, Wis., 1955–1977, vol. IV, pp. 140–64.
78. Hill, *History of Engineering*, p. 172.
79. Fedden and Thomson, *Crusader Castles*, p. 84. (The authors do not identify the "Muslim writer.")
80. Francesco Gabrielli, *Arab Historians of the Crusades*, trans. E. J. Costello, Berkeley, Calif., 1969, pp. 318–19.
81. Paul Gille, "Construction and Building," in Daumas, II, pp. 573–74.
82. Contamine, *War in the Middle Ages*, p. 113.
83. Joseph Needham, "China's Trebuchets, Manned and Counterweighted," in Hall and West, *Pre-modern Technology*, pp. 107–19.
84. Contamine, *War in the Middle Ages*, pp. 103–4; Donald Hill, "Trebuchets," *Viator* 4 (1973), p. 110.
85. Contamine, *War in the Middle Ages*, p. 103.
86. Ibid., p. 105; White, *Medieval Technology and Social Change*, pp. 102–3.
87. Bertrand Gille, "The Assembling of Raw Materials," in Daumas, I, pp. 515–16.
88. Contamine, *War in the Middle Ages*, p. 71.
89. Ibid., pp. 71–72.
90. Robert S. Lopez, "The Evolution of Land Transport in the Middle Ages," *Past and Present* 9 (1956), p. 19.
91. Lopez, *Commercial Revolution*, p. 158.
92. Frank M. Stenton, "The Road System of Medieval England," *Economic History Review* 7 (1936), p. 3.
93. Ibid., p. 6.
94. White, *Medieval Religion and Technology*, p. 287.
95. For a summary of evidence on the origin of the pivoted axle, see Leighton, *Transport and Communication*, pp. 118–21; for horseshoes; pp. 106–8.
96. Boyer, *Medieval French Bridges*, p. 161.
97. Ibid., p. 40.
98. Ibid., p. 54, quoting John Mundy, "Charity and Social Work in Toulouse, 1100–1250," *Traditio* 22 (1966), p. 205.

99. J. Gies, *Bridges and Men*, pp. 28–31.

100. Boyer, *Medieval French Bridges*, pp. 127–28.

101. J. Gies, *Bridges and Men*, pp. 34–41.

102. Boyer, *Medieval French Bridges*, pp. 144–45.

103. Ibid., p. 125.

104. Bertrand Gille, "Toward a Technological Evolution," in Daumas, I, p. 426.

105. Eugene H. Byrne, *Genoese Shipping in the Twelfth and Thirteenth Centuries*, New York, 1970 (first pub. in 1930), pp. 6–7; Unger, *The Ship in the Medieval Economy*, pp. 140–41; Unger, "Warships," p. 243; Frederic C. Lane, *Venetian Ships and Shipbuilders of the Renaissance*, New York, 1979 (first pub. in 1934), pp. 37, 106, 245.

106. Unger, "Warships," p. 245.

107. Unger, *The Ship in the Medieval Economy*, p. 145.

108. Lane, *Venetian Ships and Shipbuilders*, pp. 207–19.

109. Needham, *Science and Civilization*, IV, pt. 1, pp. 330–31.

110. Neckam, *De naturis rerum*, p. 183.

111. Lethbridge, "Shipbuilding," in Singer, II, pp. 583–85; Barbara M. Kreutz, "Mediterranean Contributions to the Medieval Mariner's Compass," *Technology and Culture* 14 (1973), pp. 367–83.

112. Mokyr, *Lever of Riches*, p. 46. "The side rudder has often, through ignorance, been condemned as inefficient. Quite the contrary; it is not a whit inferior in performance to the stern rudder (which replaced it only by offering advantages of another kind.)" (Casson, *Ships and Seamanship*, p. 224.)

113. Pryor, *Geography, Technology, and War*, pp. 30–31.

114. Ibid., p. 36.

115. Ibid., p. 35.

116. Ibid., p. 38.

117. Richard C. Dales, *The Scientific Achievement of the Middle Ages*, Philadelphia, 1973, p. 34.

118. Stephenson, "Medieval Tinkers," p. 39.

119. Ibid., p. 40.

120. Tina Stiefel, "'Impious Men': Twelfth-century Attempts to Apply Dialectic to the World of Nature," in Pamela O. Long, ed., *Science and Technology in Medieval Society*, New York, 1985, p. 188.

121. Dales, *Scientific Achievement*, p. 61.

122. Ibid., p. 125.

123. Lopez, "Still Another Renaissance?" p. 9.

124. John Kirtland Wright, *The Geographical Lore of the Time of the Crusades: A Study in the History of Medieval Science and Tradition in Western Europe*, New York, 1965, (first pub. in 1925), p. 81; Atiya, *Crusade, Commerce, and Culture*, pp. 230–31.

125. Cited in Sayed Jafar Mahmud, *Metal Technology in Medieval India*, Delhi, 1988, p. 13.

126. Forbes and Dijksterhuis, p. 9.

127. E. J. Holmyard, "Alchemical Equipment," in Singer, II, pp. 739–41.

128. E. J. Holmyard, *Alchemy*, Harmondsworth, 1968 (first pub. in 1957), pp. 45–53.

129. Husa, *Traditional Crafts and Skills*, p. 108.

130. Dales, *Scientific Achievement*, p. 37.

131. Stiefel, "'Impious Men,'" p. 196.

132. A seventh-century bishop of Noyon felt it necessary to forbid the practice of addressing the sun and moon as "Lords." (Stephen C. McCluskey, "Gregory of Tours, Monastic Timekeeping, and Early Christian Attitudes to Astronomy," in *Isis* 81 [1990], p. 13.)

133. Ovitt, *Restoration of Perfection*, pp. 44–45.

6: THE HIGH MIDDLE AGES: 1200–1400

1. Christopher Dyer, *Standards of Living in the Later Middle Ages*, Cambridge, 1989, pp. 250–60.

2. Bernard Lewis, *The Muslim Discovery of Europe*, New York, 1982, p. 25.

3. J. R. S. Phillips, *The Medieval Expansion of Europe*, Oxford, 1988, p. 105.

4. Ibid., p. 155.

5. Lynn White quotes Ibn Sa'id on Muslims in Spain in the thirteenth century: "Very often the Andalusian princes and warriors take the neighboring Christians as models for their equipment. Their arms are identical . . . their pennons, their saddles. Similar also is their mode of fighting with bucklers and long lances. They use neither the mace nor the bow of the Arabs, but employ Frankish crossbows for sieges and . . . encounters." ("The Crusades and the Technological Thrust of the West," in White, *Medieval Religion and Technology*, p. 281.)

6. Heaton, *Economic History*, p. 129; Husa, *Traditional Crafts and Skills*, pp. 162–63.

7. Heaton, *Economic History*, p. 154; Philippe Dollinger, *La Hanse, XIIe–XVIIe siècles*, Paris, 1964; M. M. Postan, "The Trade of Medieval Europe: the North," in *The Cambridge Economic History of Europe*, vol. II, *Trade and Industry in the Middle Ages*, ed. M. M. Postan and E. E. Rich, Cambridge, 1952, pp. 223–32.

8. Faye Marie Getz, "Black Death and Silver Lining: Meaning, Continuity, and Revolutionary Change in Histories of Medieval Plague," *Journal of the History of Biology* 24 (1991), pp. 265–89.

9. F. and J. Gies, *Marriage and the Family*, pp. 223–24.

10. Lopez, *Commercial Revolution*, p. 72.

11. Duby, *Rural Economy*, pp. 88–89.

12. *Walter of Henley's Husbandry, Together with an Anonymous Husbandry, Seneschaucie, etc.*, ed. E. Lamond, London, 1890.

13. M. M. Postan, *The Medieval Economy and Society: An Economic History of Britain, 1100–1500*, Berkeley, Calif., 1972, p. 101.

14. Duby, *Rural Economy*, p. 36.

15. *Walter of Henley's Husbandry*, pp. 19, 29.

16. *The Estate Book of Henry de Bray, Northamptonshire, c. 1289–1340*, ed. D. Willis, Camden Society 3rd ser., 27 (1916), pp. xxiv–xxvii; R. A. L. Smith, "The Benedictine Contribution to Medieval Agriculture," in Smith, *Collected Papers*, London, 1947, pp. 109–10.

17. Bechmann, *Trees and Man*, p. 143. In fourteenth-century France, forests had declined from 30 million hectares under Charlemagne to about 13 million; in England forest area fell from 15 percent of total land area at the time of Domesday to 10 percent in 1350.

18. *Elton Manorial Records, 1279–1351*, ed. S. C. Ratcliff, trans. D. M. Gregory, Cambridge, 1946, pp. 351, 359, 361.

19. Zvi Razi, *Life, Marriage, and Death in a Medieval Parish: Economy, Society, and Demography in Halesowen, 1270–1400*, Cambridge, 1980.

20. Bechmann, *Trees and Man*, pp. 110, 154.

21. Duby, *Rural Economy*, p. 334.

22. Ibid., p. 357.

23. Harry Miskimin, *The Economy of Early Renaissance Europe, 1300–1460*, Englewood Cliffs, N.J., 1969, pp. 32–72.

24. Frances and Joseph Gies, *Women in the Middle Ages*, New York, 1978, pp. 168–69.

25. Georges Espinas, *Les Origines du capitalisme, I: Sire Jehan Boinebroke, patricien et drapier douaisien*, Lille, 1933.

26. Henri Pirenne, *Economic and Social History of Medieval Europe*, trans. I. E. Clegg, New York, 1937, p. 187; Endrei, *Evolution des techniques*, pp. 91–92.

27. Pacey, *Technology in World Civilization*, pp. 23–24.

28. Bertrand Gille, "Machines," in Singer, II, p. 644; Bertrand Gille, "Techniques of Assembly," in Daumas, II, p. 92; Endrei, *Evolution des techniques*, pp. 52–55, 85–90.

29. Penelope Walton, "Textiles," in Blair and Ramsay, *Medieval Industries*, p. 326.

30. Endrei, *Evolution des techniques*, pp. 163–64.

31. Ibid., pp. 50–51.

32. Ibid., pp. 59–61.

33. Ibid., p. 117.

34. Lopez, *Commercial Revolution*, pp. 134–35.

35. Needham, *Science and Civilization*, vol. IV, pt. 2, pp. 404–5; B. Gille, "Assembling of Raw Materials," in Daumas, I, p. 512.

36. Patterson, "Spinning and Weaving," in Singer, II, p. 207.

37. Ibid., p. 196.

38. White, *Medieval Religion and Technology*, pp. 274–75.

39. Mazzaoui, *Italian Cotton Industry*, p. 60.

40. Ibid., pp. 62, 66–68.

41. Ibid., pp. 79–80.

42. Ibid., pp. 88–90.

43. Ibid., pp. 93–94.

44. Ibid., p. 95.

45. Ibid., pp. 98–99.

46. Ibid., pp. 51–53.

47. B. Gille, "Problems of Power and Mechanization," in Daumas, I, p. 457.

48. B. Gille, "Techniques of Assembly," in Daumas, II, p. 98; Maurice Audin, "Printing—Origins and Early Development," in Daumas, II, pp. 629–30; Derry and Williams, p. 234.

49. Pacey, *Technology in World Civilization*, pp. 42–43.

50. T. Reynolds, *Stronger Than a Hundred Men*, pp. 84–85.

51. Elizabeth Eisenstein, *The Printing Press as an Agent of Change*, Cambridge, 1979, vol. I, pp. 12–13.

52. J. and F. Gies, *Merchants and Moneymen*, pp. 85–86.

53. Ibid., pp. 140–43, 86. The change in business organization can be seen in the records of the Alberti company of Florence, which in 1307 employed fourteen factors and two years later twenty, all

under contracts specifying salary, duties, and obligations. Ownership and management of the company remained firmly in the hands of three Alberti brothers, succeeded in time by their sons.

54. Ibid., pp. 86–87.
55. Bertrand Gille, "The Technology and Civilization of the Medieval West," in Daumas, I, pp. 568–69; Lopez and Raymond, *Medieval Trade*, pp. 359–77; Edward Peragallo, *Origin and Evolution of Double-Entry Bookkeeping: A Study of Italian Practice Since the Fourteenth Century*, New York, 1938, pp. 3–16.
56. Peragallo, *Double-Entry Bookkeeping*, pp. 22–29; Iris Origo, *The Merchant of Prato: Francesco di Marco Datini, 1335–1410*, Boston, 1986 (first pub. in 1957); Enrico Bensa, *Francesco di Marco da Prato: notizie e documenti sulla mercatura italiana del secolo XIV*, Milan, 1928.
57. Lopez and Raymond, *Medieval Trade*, pp. 230–32; J. and F. Gies, *Merchants and Moneymen*, pp. 191–93.
58. George Huppert, *After the Black Death: A Social History of Early Modern Europe*, Bloomington, Ind., 1986, p. 18.
59. Dyer, *Standards of Living*, p. 189.
60. Goodchild, "Roads and Land Travel," in Singer, II, p. 533.
61. Bechmann, *Trees and Man*, p. 153.
62. Cipolla, *Before the Industrial Revolution*, p. 289.
63. Forbes, "Roads and Land Travel," in Singer, II, p. 533; Derry and Williams, pp. 177–78.
64. Dyer, *Standards of Living*, p. 209. Attempts were also made to limit indiscriminate disposal of wastewater and waste matter by dumping in the street. The "Customs of Avignon" of 1243 decreed that "no one shall have a water pipe or pipes emptying into the public street through which water flows out onto the street . . . with the exception of rain water or well water . . . Likewise, we decree that no one shall throw water onto the street, nor any steaming liquid, nor chaff, nor the refuse of grapes, nor human filth, nor bath water, nor indeed any dirt . . . And he who commits this offense, be he head of the family or not, shall pay a fine of two shillings." (John H. Mundy and Peter Riesenberg, *The Medieval Town*, Princeton, N.J., 1958, pp. 157–58.)
65. Forbes, "Hydraulic Engineering and Sanitation," in Singer, II, pp. 689–90.
66. *Vita St. Bernardi*, in Migne, *Patrologia Latina*, vol. 185, col. 570–72, cited in B. Gille, "Machines," in Singer, II, p. 650.

67. Goodchild, "Roads and Land Travel," in Singer, II, p. 533.
68. Dyer, *Standards of Living*, p. 191.
69. Goodchild, "Roads and Land Travel," in Singer, II, p. 533.
70. Gimpel, *Medieval Machine*, p. 91. In Germany and England, the shutdown took place later, in association with the Reformation.
71. Huppert, *After the Black Death*, p. 53. According to Robert S. Lopez (*The Birth of Europe*, New York, 1967, p. 261), in Florence in 1336 there were "between 8,000 and 10,000 school boys learning to read, more than 1,000 studying mathematics and 550 to 600 grappling with literature and philosophy."
72. P. Gille, "Construction and Building," in Daumas, II, p. 526.
73. Dyer, *Standards of Living*, p. 203.
74. Paul Gille, "Hydraulic Works and Water-Supply Systems," in Daumas, II, p. 527.
75. Lopez, *Commercial Revolution*, p. 126. Mundy and Riesenberg, in *The Medieval Town* (p. 37), break down the 130 crafts of the Paris tax list of 1292 into several groupings: 18 crafts in "alimentation and consumption goods such as firewood," 36 in clothing and personal furnishings, 22 in metallurgy, 22 in textiles and leather, 10 in house furniture, 5 in building and monumental arts, 3 in medicine and sanitation, and 15 in "divers specialities, including banking, brokerage, and bookmaking."
76. Dyer, *Standards of Living*, pp. 192, 210.
77. W. H. Auden, "In Time of War," in *The Selected Poetry of W. H. Auden*, New York, 1945, p. 340.
78. John Hooper Harvey, *The Gothic World, 1100–1600: A Survey of Architecture and Art*, New York, 1969 (first pub. in 1950), p. 7.
79. J. and F. Gies, *Life in a Medieval City*, p. 139.
80. Harvey, *Gothic World*, p. 27.
81. *Hamlet*, III, iv.
82. Harvey, *Gothic World*, pp. 14–17, 39–52, 157–60. According to Lon R. Shelby ("The Geometrical Knowledge of Mediaeval Master Masons," *Speculum* 47 [1972], pp. 397–98), master masons learned their geometry not in the cathedral schools or the universities but on the job, as part of the esoteric knowledge passed from master to apprentice and from father to son.
83. Lon R. Shelby, "The Role of the Master Mason in Mediaeval English Building," *Speculum* 39 (1964), pp. 387–403.
84. Pacey, *Maze of Ingenuity*, pp. 51–52.
85. Harvey, *Mediaeval Craftsmen*, p. 64.

86. Pacey, *Maze of Ingenuity*, p. 47.

87. Ibid., p. 9.

88. Shelby, "Role of the Master Mason," p. 399.

89. Ibid., p. 400.

90. Harvey, *Gothic World*, pp. 50–52.

91. Ibid., p. 47.

92. Pacey, *Maze of Ingenuity*, p. 48.

93. Gimpel, *Medieval Machine*, p. 141.

94. White, *Medieval Technology and Social Change*, p. 118.

95. B. Gille, "Problems of Power and Mechanization," in Daumas, I, p. 448.

96. Robert Mark and Huang Yun-Sheng, "High Gothic Structural Development: The Pinnacles of Reims Cathedral," in Long, *Science and Technology*, p. 127.

97. Forbes, *Man the Maker*, p. 117.

98. T. Reynolds, *Stronger Than a Hundred Men*, p. 64.

99. Bertrand Gille, "The Growth of Mechanization," in Daumas, II, p. 53.

100. Geddes, "Iron," p. 74.

101. B. Gille, "Transformation of Raw Materials," in Daumas, I, pp. 494–95.

102. Gale, *Iron and Steel*, p. 14; Forbes, *Man the Maker*, pp. 117–18; Bromehead, "Mining and Quarrying," in Singer, II, p. 74.

103. John Spencer, "Filarete's Description of a Fifteenth Century Iron Smelter at Ferrière," *Technology and Culture* 4 (1963), p. 202.

104. Gale, *Iron and Steel*, p. 9.

105. Aitchison, *History of Metals*, vol. I, pp. 246, 258.

106. "A Complaint Against the Blacksmiths," in *English Historical Documents, 1327–1485*, ed. A. R. Myers, London, 1969, p. 1055.

107. Geddes, "Iron," pp. 174–75.

108. A. R. Hall, "Military Technology," in Singer, II, p. 723. One explanation advanced for the failure of the longbow to diffuse is that the weapon required considerable strength and skill; the crossbow, technically far more complex, was easier to employ effectively.

109. Rosemary Ascherl, "The Technology of Chivalry in Reality and Romance," in Howell Chickering and Thomas Seiler, eds., *The Study of Chivalry: Resources and Approaches*, Kalamazoo, Mich., 1988.

110. A. R. Hall, "Guido's Texaurus," in Hall and West, *Pre-modern*

Technology, pp. 11–52; Bertrand Gille, *The Renaissance Engineers*, London, 1966, pp. 28–29.

111. Needham, *Science and Civilization*, vol. V, pt. 7, pp. 48–49, combining passages from Roger Bacon's *Opus maius* and *Opus tertium*.

112. Ibid., p. 49.

113. Ibid., pp. 570–79; Pacey, *Technology in World Civilization*, pp. 47–48.

114. Pacey, *Technology in World Civilization*, p. 49; White, *Medieval Religion and Technology*, p. 225.

115. Carlo Cipolla, *European Culture and Overseas Expansion*, London, 1970, p. 115 (footnote).

116. Contamine, *War in the Middle Ages*, pp. 139–40; Cipolla, *European Culture*, p. 36 (footnote).

117. Pacey, *Technology in World Civilization*, p. 54.

118. Petrarch, *De remediis*, bk. 1, dialogue 99, cited in Cipolla, *European Culture*, p. 35.

119. Contamine, *War in the Middle Ages*, p. 145.

120. Ibid., p. 196.

121. Ibid., p. 144.

122. Cipolla, *European Culture*, p. 36.

123. Hill, *History of Engineering*, p. 245.

124. D. S. L. Cardwell, *Turning Points in Western Technology*, New York, 1972, p. 14.

125. Derek J. de Solla Price, "Automata and the Origins of Mechanism and Mechanistic Philosophy," *Technology and Culture* 5 (1964), p. 18; Pacey, *Maze of Ingenuity*, pp. 38–39; Needham, *Science and Civilization*, vol. IV, pt. 2, p. 441.

126. Hill, *History of Engineering*, p. 223.

127. B. Gille, "Technology and Civilization of the Medieval West," in Daumas, I, p. 568. The earliest European text reference to the escapement is in Richard of Wallingford's *Tractatus horologii*; the device it describes is somewhat different from the verge and foliot: two spur gears mounted on a common axle with their teeth set out of phase; between them an anchor-shaped pallet rotates, its ends alternately catching and releasing the projecting teeth on either gear. Bert S. Hall believes that it may have derived directly from Villard de Honnecourt's sketched device. See Richard of Wallingford, *Tractatus horologii astronomici*, 3 vols., Oxford, 1976.

128. Usher, *History of Mechanical Inventions*, pp. 195–96. The word "clock" (French *cloche*, German *Glocke*) was probably used for the bells that rang monastic hours before the mechanical clock appeared on the scene. The English expression "o'clock" may have distinguished equal-hour clock time from elastic-hour seasonal time.

129. Cipolla, *European Culture*, p. 115.

130. Ibid., p. 116.

131. Hill, *History of Engineering*, pp. 243–44; Pierre Mesnage, "The Building of Clocks," in Daumas, II, pp. 283–84.

132. Usher, *History of Mechanical Inventions*, p. 196; Cipolla, *European Culture*, pp. 116–17.

133. Ibid., pp. 121–22; Price, "Automata," p. 18.

134. Usher, *History of Mechanical Inventions*, p. 197.

135. Jacques Le Goff, *Time, Work, and Culture in the Middle Ages*, trans. Arthur Goldhammer, Chicago, 1980, p. 49.

136. Cipolla, *European Culture*, pp. 119–20.

137. Le Goff, *Time, Work, and Culture*, p. 35.

138. Cipolla, *European Culture*, p. 119.

139. Harvey, *Mediaeval Craftsmen*, pp. 15–16.

140. Derry and Williams, p. 177.

141. Goodchild, "Roads and Land Travel," in Singer, II, p. 526.

142. Boyer, *Medieval French Bridges*, pp. 6–7; Lopez, *Commercial Revolution*, p. 142; Lopez, "Evolution of Land Transport," pp. 27–28.

143. Boyer, *Medieval French Bridges*, p. 105.

144. Ibid., p. 63.

145. Ibid., p. 168.

146. Ibid., p. 143 (*The Sketchbook of Villard de Honnecourt*, ed. Theodore Bowie, Bloomington, Ind., 1959, p. 130, plate 60).

147. Boyer, *Medieval French Bridges*, p. 156.

148. J. Gies, *Bridges and Men*, pp. 53–54.

149. Ibid., pp. 102–4; Hill, *History of Engineering*, pp. 70–72.

150. Marjorie Nice Boyer, "Medieval Suspended Carriages," *Speculum* 34 (1959), pp. 361–65.

151. White, *Medieval Religion and Technology*, p. 110.

152. Bertrand Gille, "The Fifteenth and Sixteenth Centuries in the Western World," in Daumas, II, pp. 18, 35.

153. Paul Gille, "Land and Water Transportation," in Daumas, II, p. 348.

154. Goodchild, "Roads and Land Travel," in Singer, II, p. 527.

155. B. Gille, "Machines," in Singer, II, p. 639.
156. Stenton, "Road System of Medieval England," p. 17.
157. Marjorie Nice Boyer, "A Day's Journey in Medieval France," *Speculum* 26 (1951), pp. 597–98.
158. Lopez, *Commercial Revolution*, p. 108.
159. Stenton, "Road System of Medieval England," p. 18.
160. Derry and Williams, p. 179; Forbes and Dijksterhuis, p. 142.
161. Lane, *Venetian Ships and Shipbuilders*, p. 16.
162. Ibid., pp. 176–78.
163. Unger, *The Ship in the Medieval Economy*, p. 183.
164. Ibid., p. 186.
165. Ibid., p. 185.
166. Lane, *Venetian Ships and Shipbuilders*, pp. 29–30.
167. Unger, *The Ship in the Medieval Economy*, pp. 215–16; Phillips, *Medieval Expansion of Europe*, pp. 218–19; Frederic C. Lane, "The Economic Meaning of the Invention of the Compass," *American Historical Review* 68 (1963), pp. 605–6.
168. Pryor, *Geography, Technology, and War*, p. 54.
169. Lane, "Economic Meaning of the Compass," pp. 608–9.
170. Ibid., pp. 608–10.
171. Forbes and Dijksterhuis, p. 143.
172. Bertrand Gille, "Transportation," in Daumas, II, p. 40.
173. Joseph and Frances Gies, *Leonard of Pisa and the New Mathematics of the Middle Ages*, New York, 1969, p. 58. An edition of *Liber abaci* (in the original Latin) was published in Rome, in *Scritti di Leonardo Pisano*, edited by Baldassarre Boncompagni (1857–1864). Leonardo's *Liber quadratorum* (Book of square numbers) was translated into French in 1952 by Paul Ver Eecke, as *Le Livre des nombres carrés*.
174. J. and F. Gies, *Leonard of Pisa*, pp. 77–84.
175. Ibid., pp. 62–63.
176. James Gairdner, ed., *The Paston Letters, 1422–1509*, London, 1904, vol. VI, p. 22.
177. Singer, "Epilogue," in Singer, II, p. 767.
178. Crombie, *Medieval and Early Modern Science*, vol. I, p. 97.
179. Stiefel, "'Impious Men,'" p. 212.
180. Dales, *Scientific Achievement*, pp. 63–70.
181. B. Gille, "Techniques of Assembly," in Daumas, II, p. 97.
182. Dales, *Scientific Achievement*, p. 61.
183. Haskins, *Renaissance of the Twelfth Century*, p. 310.

184. Nicholas H. Steneck, "The Relevance of the Middle Ages to the History of Science and Technology," in Long, *Science and Technology*, p. 22.
185. Forbes and Dijksterhuis, p. 97.
186. Ibid., p. 98.
187. Benton, "Ideas of Order," p. 22; Richard C. Dales, "Robert Grosseteste's View on Astrology," *Medieval Studies* 29 (1967), pp. 357–63.
188. Chaucer, *Canterbury Tales*, trans. Neville Coghill, Baltimore, 1952, p. 28 (Prologue, II, lines 411ff.).
189. Crombie, *Medieval and Early Modern Science*, I, p. 227.
190. Ibid., p. 96.
191. B. Gille, "Technology and Civilization of the Medieval West," in Daumas, I, p. 571.
192. White, *Medieval Religion and Technology*, p. 39.
193. Haskins, *Renaissance of the Twelfth Century*, p. 333.
194. Crombie, *Medieval and Early Modern Science*, I, p. 182.
195. Samuel Eliot Morison, *The European Discovery of America: The Southern Voyages*, A.D. 1492–1616, New York, 1974, pp. 27, 39.
196. Charles Homer Haskins, *Studies in Mediaeval Culture*, Oxford, 1929, p. 25.

7: LEONARDO AND COLUMBUS: THE END OF THE MIDDLE AGES

1. Lopez, "Still Another Renaissance?" p. 2.
2. Haskins, *Renaissance of the Twelfth Century*.
3. Bert S. Hall, "The New Leonardo," *Isis* 67 (1976), p. 475. In his review of *Leonardo da Vinci, the Madrid Codices*, Hall quotes an "offhand remark" of Ladislao Reti, the editor of the Codices, on Leonardo's two métiers: "At last people will start believing me when I tell them that Leonardo da Vinci was an engineer who occasionally painted a picture when he was broke" (p. 475).
4. William of Malmesbury, *Chronicle of the Kings of England*, pp. 251–52; Lynn White, "Eilmer of Malmesbury, an Eleventh-Century Aviator: A Case Study of Technological Innovation, Its Context and Tradition," *Technology and Culture* 2 (1961), pp. 97–111.

5. Robert Brun, "Notes sur le commerce des objets d'art en France et principalement à Avignon à la fin du XIVe siècle," *Bibliothèque de l'Ecole des Chartes* 95 (1934), pp. 327–46.

6. Phillips, *Medieval Expansion of Europe*, p. 154.

7. Audin, "Printing," in Daumas, II, pp. 639–40.

8. Ibid., pp. 622–23; B. Gille, "Techniques of Assembly," in Daumas, II, pp. 99–101; Derry and Williams, p. 235.

9. Audin, "Printing," in Daumas, II, pp. 632–34.

10. A. Rupert Hall, "Early Modern Technology to 1600," in Melvin Kranzberg and Carroll W. Pursell, Jr., eds., *Technology in Western Civilization*, vol. I, *The Emergence of Modern Industrial Society, Earliest Times to 1900*, New York, 1967, p. 101.

11. Audin, "Printing," in Daumas, II, p. 638.

12. B. Gille, "Techniques of Assembly," in Daumas, II, p. 100.

13. Ibid., pp. 99–100.

14. Ibid., p. 101; Derry and Williams, p. 239.

15. Derry and Williams, p. 239.

16. Ibid., p. 237.

17. Ibid., p. 236; Hall, "Early Modern Technology," p. 101; Forbes and Dijksterhuis, p. 145.

18. *Cymbeline*, I, i.

19. Derry and Williams, p. 236; B. Gille, "Techniques of Assembly," in Daumas, II, p. 101; Audin, "Printing," in Daumas, II, pp. 646–47.

20. Elizabeth E. Eisenstein, *The Printing Press as an Agent of Change*, Cambridge, 1979, vol. I, p. 54.

21. Ivor B. Hart, *The World of Leonardo da Vinci, Man of Science, Engineer, and Dreamer of Flight*, New York, 1962, p. 61.

22. Eisenstein, *Printing Press*, I, p. 46.

23. Audin, "Printing," in Daumas, II, pp. 643–44.

24. Eisenstein, *Printing Press*, I, p. 55.

25. Derry and Williams, p. 240.

26. Hart, *World of Leonardo da Vinci*, p. 62.

27. Samuel Eliot Morison, *The European Discovery of America: The Northern Voyages, A.D. 500–1600*, New York, 1971, p. 105.

28. A. C. Crombie, *Medieval and Early Modern Science*, vol. II, *Science in the Later Middle Ages and Early Modern Times, XIII–XVII Centuries*, New York, 1959 (first pub. in 1952), pp. 111–12.

29. Derry and Williams, p. 235.

30. Contamine, *War in the Middle Ages*, p. 145.

31. Jules Quicherat, ed., *Procès de condamnation et de réhabilitation de Jeanne d'Arc, dite la pucelle*, Paris, 1841–1849, vol. III, p. 211, cited in Frances Gies, *Joan of Arc: The Legend, the Reality*, New York, 1981, p. 77.

32. Ritchie Calder, *Leonardo and the Age of the Eye*, New York, 1970, p. 94.

33. Contamine, *War in the Middle Ages*, pp. 142–43; Forbes and Dijksterhuis, p. 140.

34. Cipolla, *European Culture*, p. 75.

35. Paul Gille, "Military Techniques," in Daumas, II, pp. 474–75.

36. Bertrand Gille, "Military Techniques," in Daumas, II, p. 107.

37. P. Gille, "Military Techniques," in Daumas, II, p. 477.

38. B. Gille, "Military Techniques," in Daumas, II, p. 103.

39. Cipolla, *European Culture*, p. 38.

40. B. Gille, "Military Techniques," in Daumas, II, p. 114.

41. Needham, *Science and Civilization*, vol. V, pt. 7, pp. 16–18.

42. B. Gille, *Renaissance Engineers*, p. 58.

43. Honoré Bonet, *The Tree of Battles*, trans. G. W. Coopland, Cambridge, Mass., 1949.

44. A. E. Housman, "Epitaph on an Army of Mercenaries," verse 37, *Last Poems* (1922).

45. B. Gille, *Renaissance Engineers*, p. 63.

46. Ibid., p. 12.

47. Maurice Daumas and André Garanger, "Industrial Mechanization," in Daumas, II, p. 271.

48. Calder, *Leonardo*, p. 255, quoting MS. *Trattato della pittura*, in the Vatican Library.

49. Bert S. Hall, "Der Meister sol auch kennen schreiben und lesen: Writings About Technology, ca. 1400–1600 A.D., and Their Cultural Implications," in Denise Schmant-Besserat, ed., *Early Technologies*, Malibu, Calif., 1979, p. 48.

50. B. Gille, *Renaissance Engineers*, p. 92; Calder, *Leonardo*, p. 51.

51. Hall, "Der Meister," p. 53.

52. Bertrand Gille, "Engineers and Technicians of the Renaissance," in Daumas, II, p. 24.

53. Frank D. Prager and Gustina Scaglia, *Mariano Taccola and His Book, "De ingeniis,"* Cambridge, 1972; B. Gille, *Renaissance Engineers*, pp. 83–86; B. Gille, "Engineers and Technicians of the Renaissance," in Daumas, II, p. 25.

54. Hill, *History of Engineering*, p. 73.

55. B. Gille, *Renaissance Engineers*, p. 71.
56. Ibid., p. 56; B. Gille, "Engineers and Technicians of the Renaissance," in Daumas, II, p. 23.
57. Morison, *European Discovery: The Southern Voyages*, p. 27; Phillips, *Medieval Expansion of Europe*, p. 233; Hart, *World of Leonardo da Vinci*, pp. 29–30.
58. Cited in Hart, *World of Leonardo da Vinci*, p. 142.
59. B. Gille, *Renaissance Engineers*, pp. 92–93; B. Gille, "Engineers and Technicians of the Renaissance," in Daumas, II, p. 26.
60. B. Gille, *Renaissance Engineers*, pp. 87–89; Calder, *Leonardo*, pp. 97–99; Hart, *World of Leonardo da Vinci*, p. 167; White, *Medieval Technology and Social Change*, p. 114.
61. Hart, *World of Leonardo da Vinci*, p. 139.
62. B. Gille, *Renaissance Engineers*, p. 139.
63. Ibid., pp. 106–14; B. Gille, "Engineers and Technicians of the Renaissance," in Daumas, II, p. 29.
64. B. Gille, "Engineers and Technicians of the Renaissance," in Daumas, II, pp. 43–44.
65. Ibid., pp. 49–50.
66. Hart, *World of Leonardo da Vinci*, p. 30.
67. Pacey, *Maze of Ingenuity*, p. 57.
68. Hart, *World of Leonardo da Vinci*, pp. 30–31; Charles Singer, quoted in Hart, p. 24.
69. Ibid., p. 18.
70. B. Gille, *Renaissance Engineers*, p. 138.
71. B. Gille, "Machines," in Singer, II, p. 654.
72. Calder, *Leonardo*, pp. 137–38.
73. Ibid., p. 132; B. Gille, "Machines," in Singer, II, pp. 652–56.
74. B. Gille, "Machines," in Singer, II, p. 656; Hart, *World of Leonardo da Vinci*, pp. 224–25.
75. Calder, *Leonardo*, p. 122.
76. Endrei, *Evolution des techniques*, p. 145.
77. Calder, *Leonardo*, pp. 79, 99, 92; Hart, *World of Leonardo da Vinci*, p. 273.
78. Calder, *Leonardo*, p. 197.
79. B. Gille, "Machines," in Singer, II, p. 652.
80. Ladislao Reti, "Francesco di Giorgio Martini's Treatise on Engineering and Its Plagiarists," *Technology and Culture* 4 (1963), pp. 287–98. Leonardo's own concern over plagiarism may have contributed to the obscurity into which his work sank so strangely and for so long. (Some writers have suggested that the famous

"mirror writing" in which the notebooks are composed was a precaution against plagiarism, but the conjecture seems dubious.) Italian writers were more sensitive to the problem than their German peers, who regarded texts, drawings, and ideas as public property from the moment they appeared; for lack of patent protection, Gutenberg died in poverty. (Hall, "Der Meister," p. 55.)

81. Cipolla, *Before the Industrial Revolution*, pp. 182–83.
82. B. Gille, "Engineers and Technicians of the Renaissance," in Daumas, II, p. 45.
83. Ibid., p. 42.
84. White, *Medieval Religion and Technology*, p. 49.
85. Derry and Williams, pp. 255–56.
86. T. Reynolds, *Stronger Than a Hundred Men*, p. 69.
87. Needham, *Science and Civilization*, vol. IV, pt. 2, p. 414.
88. Bertrand Gille, "Techniques of Acquisition," in Daumas, II, pp. 66–68.
89. Derry and Williams, p. 129.
90. E. M. Jope, "Vehicles and Harness," in Singer, II, p. 548; B. Gille, "Machines," in Singer, II, p. 655.
91. B. Gille, "Techniques of Acquisition," in Daumas, II, p. 76.
92. Gale, *Iron and Steel*, pp. 19–20.
93. Ibid., pp. 21–22.
94. A. R. Hall, "Early Modern Technology," p. 93; Bertrand Gille, "Transformation of Matter," in Daumas, II, pp. 76–77.
95. Calder, *Leonardo*, pp. 125–26.
96. A. R. Hall, "Early Modern Technology," p. 93.
97. Duby, *Rural Economy*, p. 302.
98. B. Gille, "Fifteenth and Sixteenth Centuries in the Western World," in Daumas, II, p. 14.
99. Endrei, *Evolution des techniques*, pp. 99–103.
100. Ibid., pp. 103–4.
101. Ibid., pp. 128–29.
102. Ibid., pp. 129–30.
103. Mazzaoui, *Italian Cotton Industry*, pp. 131–32, 138–40.
104. Ibid., pp. 142–44, 154–58, 161–62.
105. Lane, *Venetian Ships and Shipbuilders*, p. 172.
106. Prager and Scaglia, *Mariano Taccola*, pp. 92–93.
107. B. Gille, "Machines," in Singer, II, p. 657; Forbes and Dijksterhuis, p. 142; Paul Gille, "Sea and River Transportation," in Daumas, II, p. 423.
108. Calder, *Leonardo*, pp. 128–29.

109. Mesnages, "Building of Clocks," in Daumas, II, pp. 295–300.

110. Cipolla, *European Culture*, pp. 123–24.

111. Derry and Williams, p. 227; Mesnages, "Building of Clocks," in Daumas, II, pp. 298–99.

112. Usher, *History of Mechanical Inventions*, p. 209.

113. P. Gille, "Military Techniques," in Daumas, II, pp. 474–75.

114. Curt S. Gutkind, *Cosimo de' Medici, Pater Patriae, 1389–1464*, Oxford, 1938, pp. 235–36.

115. Forbes and Dijksterhuis, p. 142; Calder, *Leonardo*, p. 129.

116. Unger, *The Ship in the Medieval Economy*, p. 216.

117. Cipolla, *Before the Industrial Revolution*, p. 176.

118. Unger, *The Ship in the Medieval Economy*, p. 218.

119. Morison, *European Discovery: The Northern Voyages*, p. 115.

120. Ibid., p. 124.

121. Unger, *The Ship in the Medieval Economy*, pp. 217–21.

122. Lane, *Venetian Ships and Shipbuilders*, pp. 41–42.

123. Derry and Williams, pp. 203–4.

124. Lane, *Venetian Ships and Shipbuilders*, p. 47.

125. Phillips, *Medieval Expansion of Europe*, p. 230.

126. Morison, *European Discovery: The Southern Voyages*, p. 56.

127. Not all scholars agree about the value of medieval navigational instruments. Derek de Solla Price went so far as to assert that "no good navigator really believed instruments until about the late eighteenth century." ("Proto-Astrolabes, Proto-Clocks, and Proto-Calculators: The Point of Origin of High Mechanical Technology," in Schmant-Besserat, *Early Technologies*, pp. 47–58.)

128. Calder, *Leonardo*, p. 51.

129. Phillips, *Medieval Expansion of Europe*, p. 231.

130. Unger, *The Ship in the Medieval Economy*, p. 215.

131. Luisa Cogliati Arano, *The Medieval Health Handbook, Tacuinum sanitatis*, trans. Oscar Ratti and Adele Westbrook, New York, 1976.

132. Lopez and Raymond, *Medieval Trade*, p. 108.

133. Ibid., pp. 108–14; Francesco Balducci Pegolotti, *La pratica della mercatura*, ed. Allan Evans. Cambridge, Mass., 1936.

134. Anthony Bryer, "Europe and the Wider World to the Fifteenth Century," in Douglas Johnson, ed., *The Making of the Modern World*, vol. I, *Europe Discovers the World*, London, 1971, p. 58.

135. Cipolla, *European Culture*, pp. 99–101.

136. Morison, *European Discovery: The Northern Voyages*, p. 160.
137. B. Gille, "Techniques of Acquisition," in Daumas, II, p. 61.
138. Phillips, *Medieval Expansion of Europe*, p. 244.
139. Ibid., p. 245.
140. Jones, *History of the Vikings*, p. 308.
141. Fernand Braudel, *Civilization and Capitalism, 15th–18th Century*, vol. I, *The Structures of Everyday Life: The Limits of the Possible*, trans. Siân Reynolds, New York, 1981, p. 407.
142. Reay Tannahill, *Food in History*, New York, 1973, pp. 241–42.
143. Phillips, *Slavery from Roman Times*, p. 176.
144. Morison, *European Discovery: The Southern Voyages*, pp. 95–96.
145. Cardwell, *Turning Points*, p. 6.
146. Needham, "Chinese Priorities in Cast Iron Metallurgy," p. 402.
147. Timo Myllyntaus, "The Transfer of Electrical Technology to Finland, 1870–1930," *Technology and Culture* 32 (1991), pp. 293–94.
148. Cardwell, *Turning Points*, p. 1.
149. Cipolla, *European Culture*, p. 16.
150. Ibid., p. 18.
151. T. Reynolds, *Stronger Than a Hundred Men*, pp. 96–97.
152. B. Gille, "Machines," in Singer, II, p. 649.
153. T. Reynolds, *Stronger Than a Hundred Men*, p. 96.
154. Price, "Automata," p. 21.
155. Dales, *Scientific Achievement*, p. 176.
156. Bertrand Gille, "Evolution of the Technical Civilization," in Daumas, II, p. 135.
157. Cipolla, *Before the Industrial Revolution*, p. 112.
158. Bechmann, *Trees and Man*, p. x.
159. Cipolla, *Before the Industrial Revolution*, p. 111.
160. Lane, *Venetian Ships and Shipbuilders*, pp. 218–32.
161. Melvin Kranzberg, "Technology and History: Kranzberg's Laws," *Technology and Culture* 27 (1986), p. 545. Kranzberg explains his paradoxical First Law by elaborating: "Technology's interaction with the social ecology is such that technical developments frequently have environmental, social, and human consequences that go far beyond the immediate purposes of the technical devices and practices themselves, and the same technology can have quite different results when introduced into different contexts or under different circumstances."

BIBLIOGRAPHY

Adas, Michael. *Machines as the Measure of Men: Science, Technology, and Ideologies of Western Dominance*. Ithaca, New York, 1989.

Aitchison, Leslie. *A History of Metals*. 2 vols. London, 1960.

Arano, Luisa Cogliati. *The Medieval Health Handbook, Tacuinum sanitatis*. Translated by Oscar Ratti and Adele Westbrook. New York, 1976.

Ascherl, Rosemary. "The Technology of Chivalry in Reality and Romances." In *The Study of Chivalry: Resources and Approaches*, edited by Howell Chickering and Thomas Seiler, 263–311. Kalamazoo, Michigan, 1988.

Atiya, Aziz S. *Crusade, Commerce, and Culture*. New York, 1966 (first published in 1962).

Augustine. *City of God*. Translated by Gerald G. Walsh, Demetrius B. Zema, Grace Monahan, and Daniel Honan. New York, 1958.

Bachrach, Bernard S. "Charles Martel, Mounted Shock Combat, the Stirrup, and Feudal Origins." *Studies in Medieval and Renaissance History* 7 (1970): 47–76.

———. *Merovingian Military Organization, 481–751*. Minneapolis, 1972.

———. "The Origin of Armorican Chivalry." *Technology and Culture* 10 (1969): 166–71.

Bautier, Robert H. *The Economic Development of Medieval Europe*. London, 1971.

Bechmann, Roland. *Trees and Man: The Forest in the Middle Ages*. Translated by Katharyn Dunham. New York, 1990.

Bede. *A History of the English Church and People*. Translated by Leo Sherley-Price. Harmondsworth, 1986.

Bedini, Silvio. "The Role of Automata in the History of Technology." *Technology and Culture* 4 (1964): 24–42.

Bennett, Judith M. "The Village Ale-Wife. Women and Brewing in Fourteenth Century England." In *Women and Work in Preindustrial Europe*, edited by Barbara A. Hanawalt, 20–36. Bloomington, Indiana, 1986.

Benoit, P., and P. Braunstein, eds. *Mines, carrières, et métallurgie dans la France médiévale*. Paris, 1983.

Benton, John F. "Ideas of Order: Music, Mathematics, and Medieval Architecture." Caltech Lecture Series.

Benz, Ernest. "I fondamenti cristiani della tecnica occidentale." In Enrico Castelli, ed., *Tecnica e casistica*, 241–63. Rome, 1964.

Bivar, A. D. H. "Cavalry Equipment and Tactics on the Euphrates Frontier." *Dumbarton Oaks Papers* 26 (1979): 273–91.

———. "The Stirrup and Its Origins." *Oriental Art*, n.s. 1 (1955): 61–68.

Blair, John W., and Nigel Ramsay, eds. *English Medieval Industries*. London, 1990.

Bloch, Marc. *Land and Work in Medieval Europe: Selected Papers*. Translated by J. E. Anderson. Berkeley, California, 1967.

Boase, T. S. R. "Military Architecture in the Crusader States in Palestine and Syria." In Kenneth Setton, ed., *A History of the Crusades*, vol. IV, 140–64. Madison, Wisconsin, 1955–1977.

Boissonade, P. *Life and Work in Medieval Europe: The Evolution of Medieval Economy from the Fifth to the Fifteenth Century*. Translated by Eileen Power. New York, 1964 (first published in 1927).

Bonet, Honoré. *The Tree of Battles*. Translated by G. W. Coopland. Cambridge, Massachusetts, 1949.

Boussard, Jacques. *The Civilization of Charlemagne*. Translated by Frances Partridge. New York, 1976.

Bowles, Edmund A. "On the Origin of the Keyboard Mechanism in the Late Middle Ages." *Technology and Culture* 7 (1966): 152–62.

Boyer, Marjorie Nice. "A Day's Journey in Medieval France." *Speculum* 26 (1951): 597–608.

———. *Medieval French Bridges*. Cambridge, Massachusetts, 1976.

———. "Medieval Pivoted Axles." *Technology and Culture* 1 (1960): 128–38.

———. "Medieval Suspended Carriages." *Speculum* 34 (1959): 359–66.

———. "Rebuilding the Bridge at Albi, 1408–1410." *Technology and Culture* 10 (1969): 24–37.

———. "Roads and Rivers: Their Use and Disuse in Late Medieval France." *Medievalia et Humanistica* 13 (1960): 68–80.

———. "Water Mills: A Problem for the Bridges and Boats of Medieval France." *History of Technology* 7 (1982): 1–22.

Bradbury, Jim. *The Medieval Archer.* New York, 1985.

Braudel, Fernand. *Civilization and Capitalism, 15th–18th Century,* vol. I, *The Structures of Everyday Life: The Limits of the Possible.* Translated by Siân Reynolds. New York, 1981.

Bridbury, A. R. *Medieval English Clothmaking: An Economic Survey.* London, 1982.

Brun, Robert. "Notes sur le commerce des objets d'art en France et principalement à Avignon à la fin du XIVe siècle." *Bibliothèque de l'Ecole des Chartes* 95 (1934): 327–46.

Bryer, Anthony. "Europe and the Wider World to the Fifteenth Century." In Douglas Johnson, ed., *The Making of the Modern World,* vol. I, *Europe Discovers the World,* 50–92. London, 1971.

Byrne, Eugene H. *Genoese Shipping in the Twelfth and Thirteenth Centuries.* New York, 1970 (first published in 1930).

Calder, Ritchie. *Leonardo and the Age of the Eye.* New York, 1970.

The Cambridge Economic History of Europe, vol. II, *Trade and Industry in the Middle Ages,* edited by M. M. Postan and E. E. Rich. Cambridge, 1952. Vol. III, *Economic Organization and Policies in the Middle Ages,* edited by M. M. Postan, E. E. Rich, and Edward Miller. Cambridge, 1963.

Cardwell, D. S. L. *Turning Points in Western Technology.* New York, 1972.

Carolingian Chronicles: Royal Frankish Annals; Nithard's Histories. Translated by Bernard Walter Scholz and Barbara Rogers. Ann Arbor, 1972.

Carus-Wilson, E. M. "An Industrial Revolution of the Thirteenth Century." *Economic History Review* 12 (1941): 39–60.

Casson, Lionel. "The Grain Trade of the Hellenistic World." *Transactions and Proceedings of the American Philological Association* 85 (1954): 168–87.

———. "New Light on Ancient Rigging and Boatbuilding." *American Neptune* 24 (1964): 81–94.

———. "Odysseus' Boat (Od. V, 244–57)." *American Journal of Philology* 85 (1964): 61–94.

————. *Ships and Seamanship in the Ancient World*. Princeton, New Jersey, 1971.

The Chronicle of Theophanes. Translated by Harry Turtledove. Philadelphia, 1982.

Cipolla, Carlo. *Before the Industrial Revolution*. New York, 1980.

————. *European Culture and Overseas Expansion*. London, 1970.

Contamine, Philippe. *War in the Middle Ages*. Translated by Michael Jones. London, 1984.

Crombie, A. C. *Medieval and Early Modern Science*. Vol. I, *Science in the Middle Ages: V–XIII Centuries*. Vol. II, *Science in the Later Middle Ages and Early Modern Times: XIII–XVII Centuries*. New York, 1959 (first published in 1952).

————, ed. *Scientific Change*. New York, 1963.

Crossley, David W., ed. *Medieval Industry* (Research Report 40, Council for British Archaeology). London, 1981.

Dales, Richard C. *The Scientific Achievement of the Middle Ages*. Philadelphia, 1973.

Daniel-Rops, Henri. *Cathedral and Crusade*. London, 1956.

Daumas, Maurice, ed. *A History of Technology and Invention: Progress Through the Ages*. Vols. I and II. Translated by Eileen B. Hennessy. New York, 1970.

De Camp, L. Sprague. *The Ancient Engineers*. New York, 1963.

Derry, T. K., and Trevor I. Williams. *A Short History of Technology from the Earliest Times to A.D. 1900*. Oxford, 1960.

Dockès, Pierre. *Medieval Slavery and Liberation*. Translated by Arthur Goldhammer. Chicago, 1982.

Dollinger, Philippe. *La Hanse, XIIe–XVIIe siècles*. Paris, 1964.

Douglas, David C. *William the Conqueror*. Berkeley, California, 1967.

Drachmann, A. G. *The Mechanical Technology of Greek and Roman Antiquity*. Madison, Wisconsin, 1963.

Drew, Katherine Fischer, trans. and ed. *The Burgundian Code: Book of Constitutions or Law of Gundobad, Additional Enactments*. Philadelphia, 1972.

————. *The Laws of the Salian Franks*. Philadelphia, 1991.

————. *The Lombard Laws*. Philadelphia, 1973.

Drucker, Peter F. "The First Technological Revolution and Its Lessons." *Technology and Culture* 7 (1966): 143–51.

Duby, Georges. *The Early Growth of the European Economy: Warriors and Peasants from the Seventh to the Twelfth Century*. Translated by Howard B. Clarke. Ithaca, New York, 1974.

————. *Rural Economy and Country Life in the Medieval West*. Translated by Cynthia Postan. Columbia, South Carolina, 1968.

Dyer, Christopher. *Standards of Living in the Later Middle Ages*. Cambridge, 1989.

Einhard and Notker the Stammerer. *Two Lives of Charlemagne*. Translated by Lewis Thorpe. Harmondsworth, 1979.

Eisenstein, Elizabeth E. *The Printing Press as an Agent of Change*. 2 vols. Cambridge, 1979.

Elton Manorial Records, 1279–1351. Edited by S. C. Ratcliff, translated by D. M. Gregory. Cambridge, 1946.

Endrei, Walter. *L'Evolution des techniques du filage et du tissage du Moyen Age à la révolution industrielle*. Translated from the Hungarian by Joseph Tackacs. Paris, 1968.

Espinas, Georges. *Les Origines du capitalisme, I: Sire Jehan Boinebroke, patricien et drapier douaisien*. Lille, 1933.

The Estate Book of Henry de Bray, Northamptonshire, c. 1289–1340. Edited by D. Willis. Camden Society 3rd ser., 27 (1916).

Etz, Donald V. "The First Technological Revolution." *Technology and Culture* 8 (1967): 505–16.

Fedden, Robin R., and John Thomson. *Crusader Castles*. London, 1957.

Fernandez-Armesto, Felipe. *Before Columbus: Exploration and Colonization from the Mediterranean to the Atlantic, 1229–1492*. Philadelphia, 1987.

Finley, M. I. "Technical Innovation and Economic Progress in the Ancient World." *Economic History Review*, 2nd ser., 18 (1965): 29–45.

Foley, Vernard, Werner Soedel, John Turner, and Brian Wilhoite. "The Origin of Gearing." *History of Technology* 7 (1982): 101–29.

Forbes, R. J. *Man the Maker: A History of Technology and Engineering*. London, 1958.

Forbes, R. J., and E. J. Dijksterhuis. *A History of Science and Technology*, vol. I, *Ancient Times to the Seventeenth Century*. Harmondsworth, 1963.

Gabrielli, Francesco. *Arab Historians of the Crusades*. Translated from the Italian by E. J. Costello. Berkeley, California, 1969.

Gaier, C. "La Cavalerie lourde en Europe occidentale du XIIe au XVe siècle." *Revue internationale d'histoire militaire* 31 (1971): 385–96.

Gairdner, James, ed. *The Paston Letters, 1422–1509*. 6 vols. London, 1904.

Gale, W. K. V. *Iron and Steel*. London, 1969.

Getz, Faye Marie. "Black Death and Silver Lining: Meaning, Continuity, and Revolutionary Change in Histories of Medieval Plague." *Journal of the History of Biology* 24 (1991): 265–89.

Gibbon, Edward. *Decline and Fall of the Roman Empire*. 2 vols., n.d. (Modern Library edition). (First published in 6 vols., 1776–1788.)

Gies, Frances. *Joan of Arc: the Legend, the Reality*. New York, 1981.

———. *The Knight in History*. New York, 1984.

Gies, Frances and Joseph. *Leonard of Pisa and the New Mathematics of the Middle Ages*. New York, 1969.

———. *Life in a Medieval Castle*. New York, 1974.

———. *Life in a Medieval City*. New York, 1969.

———. *Life in a Medieval Village*. New York, 1990.

———. *Marriage and the Family in the Middle Ages*. New York, 1987.

———. *Merchants and Moneymen: The Commercial Revolution, 1000–1500*. New York, 1971.

———. *Women in the Middle Ages*. New York, 1978.

Gies, Joseph. *Bridges and Men*. New York, 1963.

Gille, Bertrand, ed. *The History of Techniques*, vol. I, *Techniques and Civilizations*. New York, 1986 (first published as *Histoire des techniques* in 1978).

———. "Le Moulin à eau, une révolution technique mediévale." *Techniques et civilisations* 3 (1954): 1–85.

———. "La Naissance du système bielle-manivelle." *Techniques et civilisations* 2 (1952): 42–46.

———. "Recherches sur les instruments du labour au Moyen Age." *Bibliothèque de l'Ecole des Chartes* 120 (1962): 5–38.

———. *The Renaissance Engineers*. London, 1966.

Gimpel, Jean. *Les Batisseurs de cathédrales*. Paris, 1964.

———. *The Medieval Machine*. New York, 1976.

Glick, Thomas F. "Levels and Levelers: Surveying Irrigation Canals in Medieval Valencia." *Technology and Culture* 9 (1968): 168–80.

Grant, Michael. *The World of Rome*. New York, 1960.

Gregory of Tours. *The History of the Franks*. Translated by Lewis Thorpe. Harmondsworth, 1974.

Gutkind, Curt S. *Cosimo de' Medici, Pater Patriae, 1389–1464*. Oxford, 1938.

Hall, Bert S. "Giovanni de' Dondi and Guido da Vigevano: Notes Toward a Typology of Medieval Technological Writings." In Madeleine Pelner Cosman and Bruce Chandler, eds., *Machaut's World: Science and Art in the Fourteenth Century*, 127–42. New York, 1978.

————. "Der Meister sol auch kennen schreiben und lesen: Writings About Technology, ca. 1400–1600 A.D., and Their Cultural Implications." In Denise Schmant-Besserat, ed., *Early Technologies*, 47–58. Malibu, California, 1979.

————. "The New Leonardo," review of Ladislao Reti, ed., *Leonardo da Vinci, the Madrid Codices* and *The Unknown Leonardo*. *Isis* 67 (1976): 463–76.

Hall, Bert S., and Delno C. West, eds. *On Pre-modern Technology and Science: A Volume of Studies in Honor of Lynn White, Jr.* Malibu, California, 1976.

Hart, Henry H. *Marco Polo, Venetian Adventurer*. Norman, Oklahoma, 1967.

Hart, Ivor B. *The World of Leonardo da Vinci, Man of Science, Engineer, and Dreamer of Flight*. New York, 1962.

Harte, N. B., and K. G. Ponting. *Cloth and Clothing in Medieval Europe: Essays in Memory of Professor E. M. Carus-Wilson*. London, 1983.

Hartenberg, Richard S., and John Schmidt, Jr. "The Egyptian Drill and the Origin of the Crank." *Technology and Culture* 10 (1969): 155–65.

Harvey, John Hooper. *The Gothic World, 1100–1600: A Survey of Architecture and Art*. New York: 1969 (first published in 1950).

————. *Mediaeval Craftsmen*. London, 1975.

Haskins, Charles Homer. *The Normans in European History*. New York, 1966 (first published in 1915).

————. *The Renaissance of the Twelfth Century*. New York, 1963 (first published in 1927).

————. *Studies in Mediaeval Culture*. Oxford, 1929.

Hatch, Mary Gies. "De gulzige Waterwolf: Medieval Dikes in Friesland." Unpublished paper.

Hawthorne, John G., and Cyril Stanley Smith. *On Divers Arts: The Treatise of Theophilus*. Chicago, 1963.

Heaton, Herbert. *Economic History of Europe*. New York, 1936.

Herlihy, David. *Opera Muliebria: Women and Work in Medieval Europe*. Philadelphia, 1990.

————, ed. *Medieval Culture and Society*. New York, 1968.

Herlihy, David, Robert S. Lopez, and Vsevolod Slessarev, eds. *Economy, Society, and Government in Medieval Italy: Essays in Memory of Robert Reynolds*. Kent, Ohio, 1969.

Hill, David, David Barrett, Keith Maude, Julia Warburton, and Margaret Worthington. "Quentovic Defined." *Antiquity* 64 (1990): 51–58.

Hill, Donald. *A History of Engineering in Classical and Medieval Times.* La Salle, Illinois, 1984.

————. "Trebuchets." *Viator* 4 (1973): 99–116.

Hitti, Philip. *History of the Arabs from the Earliest Times to the Present.* London, 1964.

Hodges, Richard, and David Whitehouse. *Mohammed, Charlemagne, and the Origins of Europe: Archaeology and the Pirenne Thesis.* Ithaca, New York, 1983.

Hoffmann, Marta. *The Warp-weighted Loom: Studies in the History and Technique of an Ancient Implement.* Oslo, 1964.

Holland, Leicester Bodine. "Traffic Ways About France in the Dark Ages (500–1150)." Ph.D. thesis, University of Pennsylvania. Allentown, 1919.

Holmes, Urban Tigner, Jr. *Daily Living in the Twelfth Century, Based on the Observations of Alexander Neckam in London and Paris.* Madison, Wisconsin, 1966 (first published in 1952).

Holmyard, E. J. *Alchemy.* Harmondsworth, 1968 (first published in 1957).

Holt, Richard. *The Mills of Medieval England.* London, 1988.

Hooper, Alfred. *Makers of Mathematics.* New York, 1948.

Hourani, George. *Arab Seafaring in the Indian Ocean in Ancient and Early Medieval Times.* Beirut, 1963.

Howell, Martha C. *Women, Production, and Patriarchy in Late Medieval Cities.* Chicago, 1986.

Huppert, George. *After the Black Death: A Social History of Early Modern Europe.* Bloomington, Indiana, 1986.

Husa, Vaclav, Josef Petrau, and Alena Surbota. *Traditional Crafts and Skills: Life and Work in Medieval and Renaissance Times.* London, 1967.

Jones, Gwyn. *A History of the Vikings.* New York, 1968.

Kilby, Kenneth. *The Cooper and His Trade.* London, 1971.

Komroff, Manuel, ed. *Contemporaries of Marco Polo.* New York, 1989 (first published in 1928).

Kranzberg, Melvin. "Technology and History: Kranzberg's Laws." *Technology and Culture* 27 (1986): 544–60.

Kranzberg, Melvin, and Carroll W. Pursell, Jr., eds. *Technology in Western Civilization,* vol. I, *The Emergence of Modern Industrial Society, Earliest Times to 1900.* New York, 1967.

Kren, Claudia, ed. *Medieval Science and Technology: A Select Annotated Bibliography.* New York, 1985.

Kreutz, Barbara M. "Mediterranean Contributions to the Medieval Mariner's Compass." *Technology and Culture* 14 (1973): 367–83.

Lambert of Ardres. *Historia comitum Ghisnensium.* In *Monumenta Germaniae historica scriptores,* edited by G. H. Pertz et al., vol. 24, ch. 127. Hanover, 1826–1913.

Landels, J. G. *Engineering in the Ancient World.* Berkeley, California, 1978.

Landes, David S. *Revolution in Time: Clocks and the Making of the Modern World.* Cambridge, Massachusetts, 1983.

Lane, Frederic C. "The Economic Meaning of the Invention of the Compass." *American Historical Review* 68 (1963): 605–17.

———. *Venetian Ships and Shipbuilders of the Renaissance.* New York, 1979 (first published in 1934).

Langford, Jerome J. *Galileo, Science, and the Church.* Ann Arbor, Michigan, 1992.

Latouche, Robert. *The Birth of Western Economy: Economic Aspects of the Dark Ages.* Translated by E. M. Wilkinson. New York, 1961.

Lavedan, Pierre. *French Architecture.* Harmondsworth, 1956.

Lefebvre des Noëttes, Richard. *L'Attelage et le cheval de selle à travers les ages.* Paris, 1931.

———. *De la marine antique à la marine moderne.* Paris, 1935.

Le Goff, Jacques. *Time, Work, and Culture in the Middle Ages.* Translated by Arthur Goldhammer. Chicago, 1980.

Leighton, Albert C. *Transport and Communication in Early Medieval Europe, A.D. 500–1100.* Newton Abbot, 1971.

Le Roy Ladurie, Emmanuel. *Times of Feast, Times of Famine: A History of Climate Since the Year 1000.* Translated by Barbara Bray. New York, 1971.

Lewis, Archibald R. "The Islamic World and the Latin West, 1350–1500." *Speculum* 65 (1990): 833–44.

———. *Nomads and Crusaders, A.D. 1000–1368.* Bloomington, Indiana, 1991.

Lewis, Bernard. *The Arabs in History.* New York, 1960 (first published in 1950).

———. *The Muslim Discovery of Europe.* New York, 1982.

Long, Pamela O., ed. *Science and Technology in Medieval Society.* New York, 1985.

Lopez, Robert S. *The Birth of Europe.* New York, 1967.

———. *The Commercial Revolution of the Middle Ages, 950–1350.* Cambridge, 1976.

———. "The Evolution of Land Transport in the Middle Ages." *Past and Present* 9 (1956): 17–29.

———. "Mohammed and Charlemagne: A Revision." *Speculum* 18 (1943): 14–38.

———. "Still Another Renaissance?" *American Historical Review* 57 (1951): 1–21.

Lopez, Robert S., and Irving W. Raymond, eds. *Medieval Trade in the Mediterranean World*. New York, 1955.

McCluskey, Stephen C. "Gregory of Tours, Monastic Timekeeping, and Early Christian Attitudes to Astronomy." *Isis* 81 (1990): 8–22.

McGrail, Sean, ed. *Maritime Celts, Frisians, and Saxons*. London, 1990.

McKitterick, Rosamond. *The Carolingians and the Written Word*. Cambridge, 1989.

Mahmud, Sayed Jafar. *Metal Technology in Medieval India*. Delhi, 1988.

Malinowski, Roman. "Ancient Mortars and Concretes: Aspects of Their Durability." *History of Technology* 7 (1982): 89–100.

Martini, Francesco di Giorgio. *Trattati di architettura, ingegneria, e arte militare*. 2 vols. Milan, 1967.

Matthies, Andrea L. "Medieval Treadwheels: Artists' Views of Building Construction." *Technology and Culture* 33 (1992): 510–47.

Mazzaoui, Maureen Fennell. *The Italian Cotton Industry in the Later Middle Ages (1100–1600)*. Cambridge, 1981.

Meursinge, John H. "Overlapping Histories of Technology." *Technology and Culture* 8 (1967): 517–18.

Milne, Gustav, and Damian Goodburn. "The Early Medieval Port of London, A.D. 700–1200." *Antiquity* 64 (1990): 629–36.

Miskimin, Harry A. *The Economy of Early Renaissance Europe, 1300–1460*. Englewood Cliffs, New Jersey, 1969.

Mokyr, Joel. *The Lever of Riches: Technological Creativity and Economic Progress*. New York, 1990.

Morison, Samuel Eliot. *The European Discovery of America: The Northern Voyages, A.D. 500–1600*. New York, 1971.

———. *The European Discovery of America: The Southern Voyages, A.D. 1492–1616*. New York, 1974.

Mundy, John H., and Peter Riesenberg. *The Medieval Town*. Princeton, New Jersey, 1958.

Myllyntaus, Timo. "The Transfer of Electrical Technology to Finland, 1870–1930." *Technology and Culture* 32 (1991): 293–317.

Neckam, Alexander. *De naturis rerum*. Translated by Thomas Wright. London, 1863.

Needham, Joseph. "Chinese Priorities in Cast Iron Metallurgy." *Technology and Culture* 5 (1964): 398–404.

————. *Science and Civilization in China*. Cambridge, 1954–. Vol. I, *Introductory Orientations*, 1954. Vol. IV, pt. 1 (with Wang Ling), *Physics*, 1962. Vol. IV, pt. 2 (with Wang Ling), *Mechanical Engineering*, 1965. Vol. V, pt. 1 (sole author, Tsien Tsuen-Hsuin), *Paper and Printing*, 1985. Vol. V, pt. 7 (with Ho Ping-Yu, Lu Gwei-Djen, and Wang Ling), *Military Technology: The Gunpowder Epic*, 1986.

North, J. D. "The Astrolabe." *Scientific American* 230 (1974): 96–106.

————, ed. and trans. *Richard of Wallingford, Tractatus horologii astronomici*. 3 vols. Oxford, 1976.

O'Leary, De Lacy. *How Greek Science Passed to the Arabs*. London, 1964.

Ovitt, George, Jr. *The Restoration of Perfection: Labor and Technology in Medieval Culture*. New Brunswick, New Jersey, 1987.

Pacey, Arnold. *The Maze of Ingenuity: Ideas and Idealism in the Development of Technology*. Cambridge, Massachusetts, 1992.

————. *Technology in World Civilization*. Cambridge, Massachusetts, 1991.

Panofsky, Erwin, ed. and trans. *Abbot Suger on the Abbey Church of St. Denis and Its Art Treasures*. Princeton, New Jersey, 1946.

Pegolotti, Francesco Balducci. *La pratica della mercatura*. Edited by Allan Evans. Cambridge, Massachusetts, 1936.

Peragallo, Edward. *Origin and Evolution of Double-Entry Bookkeeping: A Study of Italian Practice Since the Fourteenth Century*. New York, 1938.

Pevsner, Nikolaus. *An Outline of European Architecture*. Harmondsworth, 1954.

Phillips, J. R. S. *The Medieval Expansion of Europe*. Oxford, 1988.

Phillips, William D., Jr. *Slavery from Roman Times to the Early Transatlantic Trade*. Minneapolis, 1985.

Pirenne, Henri. *Economic and Social History of Medieval Europe*. Translated by I. E. Clegg. New York, 1937.

————. *Mohammed and Charlemagne*. Translated by Bernard Miall. New York, 1964 (first published in 1939).

The Politics of Aristotle. Translated by Ernest Barker. Oxford, 1948.

Postan, M. M. *The Medieval Economy and Society: An Economic History of Britain, 1100–1500*. Berkeley, California, 1972.

Pounds, Norman J. G. *Hearth and Home: A History of Material Culture*. Bloomington, Indiana, 1989.

Power, Eileen. *Medieval Women*. Edited by M. M. Postan. Cambridge, 1975.

Prager, Frank D., and Gustina Scaglia. *Mariano Taccola and His Book De ingeniis*. Cambridge, 1972.

Praus, A. A. "Mechanical Principles Involved in Primitive Tools and Those of the Machine Age." *Isis* 38 (1948): 157–69.

Price, Derek J. de Solla. "Automata and the Origins of Mechanism and Mechanistic Philosophy." *Technology and Culture* 5 (1964): 9–23.

———. "Proto-Astrolabes, Proto-Clocks, and Proto-Calculators: The Point of Origin of High Mechanical Technology." In *Early Technologies*, edited by Denise Schmant-Besserat, 61–63. Malibu, California, 1979.

Pryor, John H. *Geography, Technology, and War: Studies in the Maritime History of the Mediterranean, 649–1571*. Cambridge, 1988.

Rahtz, Philip A., and Donald Bullough. "The Parts of an Anglo-Saxon Mill." *Anglo-Saxon England* 6 (1977): 15–37.

Razi, Zvi. *Life, Marriage, and Death in a Medieval Parish: Economy, Society, and Demography in Halesowen, 1270–1400*. Cambridge, 1980.

Reti, Ladislao. "Francesco di Giorgio Martini's Treatise on Engineering and Its Plagiarists." *Technology and Culture* 4 (1963): 287–98.

Reynolds, Robert L. *Europe Emerges: Transition Toward an Industrial World-wide Society, 600–1750*. Madison, Wisconsin, 1967.

Reynolds, Terry S. *Stronger Than a Hundred Men: A History of the Vertical Water Wheel*. Baltimore, 1983.

Riché, Pierre. *Daily Life in the World of Charlemagne*. Translated by Jo Ann McNamara. Philadelphia, 1980.

Roberts, Lawrence D., ed. *Approaches to Nature in the Middle Ages*. Binghamton, New York, 1982.

Robinson, Fred C. "Medieval, the Middle Ages." *Speculum* 59 (1984): 745–56.

Roland, Alex. "Secrecy, Technology, and War: Greek Fire and the Defense of Byzantium, 678–1204." *Technology and Culture* 33 (1992): 655–79.

Runciman, Steven. *Byzantine Civilization*. New York, 1956.

Saalman, Howard. *Medieval Cities*. New York, 1968.

Sawyer, P. H., and R. H. Hilton. "Technical Determinism: The Stirrup and the Plough" (review of Lynn White, Jr., *Medieval Technology and Social Change*). *Past and Present* 24 (1963): 90–100.

Schmant-Besserat, Denise, ed. *Early Technologies*. Malibu, California, 1979.

Scritti di Leonardo Pisano. Edited by Baldassarre Boncompagni. Rome, 1857–1864.

Shapiro, Sheldon. "The Origin of the Suction Pump." *Technology and Culture* 5 (1964): 566–74.

Shelby, Lon R. "The Education of Medieval English Masons." *Medieval Studies* 33 (1970): 1–26.

———. "The Geometrical Knowledge of Mediaeval Master Masons." *Speculum* 47 (1972): 395–421.

———. "Medieval Masons' Tools: The Level and the Plumb Rule." *Technology and Culture* 2 (1961): 127–31.

———. "The Role of the Master Mason in Mediaeval English Building." *Speculum* 39 (1964): 387–403.

Singer, Charles, E. J. Holmyard, A. R. Hall, and Trevor I. Williams, eds. *A History of Technology*. Vol. II, *The Mediterranean Civilizations and the Middle Ages, c. 700 B.C. to A.D. 1500*. Oxford, 1956.

Smith, A. Mark. "Knowing Things Inside Out: The Scientific Revolution from a Medieval Perspective." *American Historical Review* 95 (1990): 726–44.

Smith, Cyril Stanley. "Granulating Iron in Filarete's Smelter." *Technology and Culture* 5 (1964): 386–90.

Smith, R. A. L. "The Benedictine Contribution to Medieval Agriculture." In Smith, *Collected Papers*, 103–16. London, 1947.

Spencer, John. "Filarete's Description of a Fifteenth Century Iron Smelter at Ferrière." *Technology and Culture* 4 (1963): 201–5.

Sprat, Thomas. *History of the Royal Society*. London, 1667.

Stenton, Frank M. "The Road System of Medieval England." *Economic History Review* 7 (1936): 1–21.

Stephenson, Carl. "In Praise of Medieval Tinkers." *Journal of Economic History* 8 (1948): 26–42.

Swetz, Frank J. *Capitalism and Arithmetic: The New Math of the 15th Century*. La Salle, Illinois, 1987.

Tannahill, Reay. *Food in History*. New York, 1973.

Tertullian. *De anima*. In J. P. Migne, ed., *Patrologia Latina*, vol. II, col. 700. Paris, 1844.

Thomas, Keith. "Work and Leisure in Pre-industrial Society." *Past and Present* 29 (1964): 50–62.

Thompson, E. A., ed. A Roman Reformer and Inventor, Being a Text of the Treatise De rebus bellicis. Oxford, 1950.

Thompson, James Westfall. "The Commerce of France in the Fifth Century." Journal of Political Economy 23 (1915): 857–87.

Toy, Sidney. The Castles of Great Britain. London, 1953.

Turgot, Anne-Robert-Jacques. On Progress, Sociology, and Economics. Edited and translated by Ronald L. Meek. Cambridge, 1973.

Turnau, Irma. "The Organization of the European Textile Industry from the Thirteenth to the Eighteenth Century." Journal of European Economic History 17 (1988): 583–602.

Turner, A. J. Astrolabes; Astrolabe-Related Instruments. Rockford, Illinois, 1985.

Unger, Richard. The Ship in the Medieval Economy, 600–1600. London, 1980.

———. "Warships and Cargo Ships in Medieval Europe." Technology and Culture 22 (1981): 233–52.

Usher, Abbott Payson. A History of Mechanical Inventions. Boston, 1959 (first published in 1929).

Villard de Honnecourt. The Sketchbook of Villard de Honnecourt. Edited by Theodore Bowie. Bloomington, Indiana, 1959.

Von Simson, Otto. The Gothic Cathedral. New York, 1956.

Walter of Henley's Husbandry, Together with an Anonymous Husbandry, Seneschaucie, etc. Edited by E. Lamond. London, 1890.

Watson, Andrew H. "The Arab Agricultural Revolution and Its Diffusion, 700–1100." Journal of Economic History 34 (1974): 8–35.

Weber, Max. The Protestant Ethic and the Spirit of Capitalism. Translated by Talcott Parsons. New York, 1958.

Weisheipl, James A. The Development of Physical Theory in the Middle Ages. Ann Arbor, Michigan, 1971.

———, ed. Albertus Magnus and the Sciences: Commemorative Essays. Toronto, 1980.

Wertime, Theodore A. "Asian Influences on European Metallurgy." Technology and Culture 5 (1964): 391–97.

White, Lynn, Jr. "Dynamo and Virgin Reconsidered." American Scholar 27 (1958): 183–94.

———. "Eilmer of Malmesbury, an Eleventh-Century Aviator: A Case Study of Technological Innovation, Its Context and Tradition." Technology and Culture 2 (1961): 97–111.

———. "The Historical Roots of Our Ecological Crisis." Science 156 (1967): 1203–7.

————. *Medieval Religion and Technology: Collected Essays*. Berkeley, California, 1978.

————. *Medieval Technology and Social Change*. London, 1962.

————. "The Study of Medieval Technology, 1924–1974, Personal Reflections." *Technology and Culture* 16 (1975): 519–30.

————. "Technology and Invention in the Middle Ages." *Speculum* 15 (1940): 141–56.

————. "What Accelerated Technological Progress in the Western Middle Ages?" In A. C. Crombie, ed., *Scientific Change*, 272–91. New York, 1963.

————, ed. *The Transformation of the Roman World: Gibbon's Problem After Two Centuries*. Berkeley, California, 1966.

Whitney, Elspeth. *Paradise Restored: The Mechanical Arts from Antiquity Through the Thirteenth Century*. Transactions of the American Philosophical Society, vol. 80. Philadelphia, 1990.

Wild, J. P. *Textile Manufacture in the Northern Roman Provinces*. Cambridge, 1970.

William of Malmesbury. *Chronicle of the Kings of England*. Edited and translated by J. A. Giles. London, 1889.

Wotton, William. *Reflections upon Ancient and Modern Learning*. London, 1894.

Wright, John Kirtland. *The Geographical Lore of the Time of the Crusades: A Study in the History of Medieval Science and Tradition in Western Europe*. New York, 1965 (first published in 1925).

Wright, Thomas, ed. *A Volume of Vocabularies*. London, 1857.

Wulff, Hans E. *The Traditional Crafts of Persia, Their Development, Technology, and Influence on Eastern and Western Civilization*. Cambridge, Massachusetts, 1966.

INDEX

References to illustrations are in italics.

Aachen, 67
abacus, 159, 225
Abelard, Peter, 160, 164
Acre, 158
Adam, 6–7, 8
Adam, William, 167
Adam of Dryburgh, 10
Adda, River, 217
Adelard of Bath, 160, 225
Adrianople, 249
Aelfric, 62
Aelred of Rievaulx, 10
Africa, 18, 277, 279, 282, 283, 284, 286
Agincourt, battle of (1415), 204
Agricola, 266
agriculture, 3, 17, 19, 23–24, 31, 37, 43, 44–49, 80, 102–103, 109–112, 158, 169–173, 268–269, 286, 290
  Bronze Age, 19
  Gaulish, 31
  Muslim, 102–103
  open-field, 3, 109–112, 268
  prehistoric, 17
  Roman, 23–24
  treatises on, 36, 44, 169–171
Ailly, Pierre d', 234, 279
Aitchison, Leslie, 64
Alaric, 281
Alberti, Leon Battista, 254, 255–256
Alberti company, 314

Albertus Magnus, 13, 229, 234, 246, 252
alchemy, 22–23, 162–164, 230, 231
alcohol, 163
Alcuin, 77
Aldric, Saint, 69
Alexander the Great, 21, 84
Alexandria, 21–22, 30–31, 100, 158, 197
Alfonso V, king of Portugal, 255
Alfonso VI, king of Castile and Leon, 106
Alfonso X (the Learned), king of Castile and Leon, 233
Aljubarrota, battle of (1385), 210
Allier, River, 217
Almagest, 161
alphabet, 21, 23
Als ship, 72
alum, 107, 181, 222
Amalfi, 156, 222
America, European discovery of, 106–107, 277–278, 284, 285–286, 306
Anna Comnena, 148
Antikythera, 296
aqueducts, 26, 69
Arabs, 40, 61, 75, 95, 100–104, 105, 113–114, 116, 122, 142–144, 145, 148, 156, 157, 158, 159, 160, 162–163, 167, 170, 207, 225, 226, 233, 237, 278, 282

arch, 19, 26, 130–132
  pointed, 130–132
  semicircular, 26
Archimedes, 22
architecture, 19–20, 26–27, 66–68,
    129–139, 167, 191, 192–200, 254,
    255, 256, 257, 271, 309
  domestic, 27, 191
  Egyptian, 19
  Gothic, 3, 129–139, 167, 192–200
  Greek, 19–20
  Roman, 26–27
  Romanesque, 66–68, 130, 131
arco, 122
Aristotle, 10, 11, 16, 21, 36, 84, 100,
    160, 161, 228–229, 236, 289
armillary sphere, 159, 231
armor, 31, 57–58, 62, 64, 107, 123, 125,
    204–205, 249–250
armorers, 126, 127, 203
arms, 31, 107, 123; see also cannon;
    crossbow; guns, hand; gunpowder
    firearms; longbow; swords
Arno, 218, 261–262
arrow loop, see meurtrière
Arte della Lana, 176
Arte di Calimala, 121–122
artesian well, 112
artist-engineers, 250–251, 252–263, 291
assarting, 111
assembly line, 271, 289
astrolabe, 23, 159, 210–211, 224, 231,
    233, 278
astrology, 22–23, 160, 162, 164,
    210–211, 230–231
  medical, 210, 230–231
astronomy, 22, 37, 159, 160, 161,
    210–211, 214, 296–297
Auden, W. H., 192
Augsburg, 266
Augustine, Saint, 6–7, 10, 76
Aulon, Jean d', 247
Ausonius, 35
automata, 22, 37, 252
Avars, 55
Averroës, 101
Avicenna, 101
Avignon, 183, 184, 235, 239, 315
Azores, 240, 283, 284, 306

Babylon, 19, 20, 214
Bacon, Francis, 162
Bacon, Roger, 13–14, 206, 207, 227,

231, 233, 234, 246, 252
Baghdad, 97, 100, 103, 106, 122, 175,
    182
Baibars, 144
ban, 115–116, 307
banking, 102, 185–186
barbarians, 1, 4, 15, 31, 39, 42
Barbegal, 35
barrel, 32, 69, 149
Bartolomaeus Anglicus, 246
baths, public, 26, 27, 190, 252
Baths of Caracalla, 130
Bayeux Tapestry, 46, 56, 57, 149
bearings, 258
Beaumaris Castle, 196
Beauvais Cathedral, 200
Bechmann, Roland, 290
Bede, 7, 68, 77
Behaim, Martin, 279
Belisarius, 70–71, 89, 113
bell founding, 128, 267, 268
Bellifortis, 250, 252, 273
bellows, 63, 87–88, 128, 163, 201
Benedict, Saint, 48, 67
Benedictine Order, 48–49, 79–80, 129,
    132
Benedictine Rule, 5, 9, 48
Bénézet, Saint, 150, 151, 153, 216
Benton, John F., 77, 230
Benz, Ernest, 5, 6
Bernard of Clairvaux, Saint, 9, 114,
    139, 188
Beukelszoon, William, 225
bill of exchange, 185
al-Biruni, 161
Black Death, 168, 172, 191, 235, 268,
    270, 283
blacksmith, 62–65, 124–125, 202–203,
    214, 215, 266, 267
blast furnace, 3, 200–202, 266, 289
Bloch, Marc, 3
Boethius, 10–11, 36, 76–77, 78, 164,
    246
Bohemia, 134, 168
Boinebroke, Jehan, 174–175, 192
Bologna, 190
Bonaventure, Saint, 13
Bonet, Honoré, 251–252
book fairs, 246
bookkeeping, double-entry, 3, 15, 169,
    184–185
books, 21, 78–79, 181–182, 190,
    241–246, 287, 289

Bougia, 225
Boulogne, 69, 70
Bourges Cathedral, 199–200
bow (in musical instruments), 102
Boyer, Marjorie Nice, 70, 149, 219
Brabant, 186
brace and bit, 66, 263
Braudel, Fernand, 283
Bray, Henry de, 171
Brazil, 284
Bremen, 107
Brendan, Saint, 106, 306
brewing, 125
brick-making, 20
bridges, 26–27, 70–71, 86, 92, 149–154,
    206–218, 255, 261, 263
  cantilever, 218
  draw, 143, 152, 261
  segmental arch, 92, 151, 218
  suspension, 92
  truss, 218, 261–262
Britain, 40–41, 43; see also England
Bronze Age, 17, 18–21, 37, 82
bronze founding, 82, 128, 267, 268
Brothers of the Bridge, 150–151, 216
Bruges, 183, 184, 186, 274
Brunelleschi, Filippo, 254, 255, 271
Bryce, James, 80
building construction, 19–20, 26–27,
    109, 129–145, 158, 192–200, 271
Bureau brothers, 248, 249
Burgos, siege of (1475), 249
Burgundy, 129, 134
Buridan, Jean, 234, 252
business techniques, 37, 102
bylaws, 111
Byzantium, 40, 50, 60–62, 71, 123, 142,
    145, 148, 157

Cabot, John, 282
Cabral, Pedro, 284
Caen, 136
Caffa, 168
caisson, 255
Calais, siege of (1346), 208
calendars, 86
Callinicus, 61
cam, 115
cambium, 109, 185
canal lock, 69, 92, 154, 221, 271,
    272–273, 287
canals, 69, 102, 117, 154, 221
Canary Islands, 234, 284

cannon, 207–209, 210, 248–249, 251,
    260–261, 267
  on shipboard, 209, 210
Canterbury, 148, 159, 188, 189
Canterbury Cathedral, 136–139, 192
Canterbury Cathedral Priory, 188–189
cantilever, see bridges, cantilever
Cape of Good Hope, 240, 283
Capitulare de villis, 48, 50
caravel, 277–278
Cardan, Jerome, 1–2, 95
Cardan suspension, 94–95
carding, 50, 175, 176–177
Cardwell, D. S. L., 210, 287, 288
Caroline minuscule, 77
Carolingian renaissance, 42, 67
carpenters, 125–126, 218
carrack, 275–276, 277
carriages, 218–219, 266
Carthusian Order, 112
carvel construction, 20, 72, 76, 276
Cassiodorus, 11, 96, 246
Casson, Lionel, 297
cast iron, 85, 87–88, 201–202
Castagno, Andrea del, 275
castellan, 59
Castile, 125, 167
Castillon, battle of (1453), 248
castles, 3, 58–60, 80, 102, 139–145,
    167, 250
  Crusader, 141–144
  motte-and-bailey, 58–60, 80, 145
  shell keep, 141
  square keep, 140–141
castles, ships', 156, 222, 233, 276
catapult, torsion, 31, 37
cathedral schools, 159, 316
Catullus, 51
Caxton, William, 245
cement, 26, 67
Ceuta, 158, 282
Ceylon, 184
chain mail, see armor
Champagne Fairs, 108–109, 121, 174,
    220, 221
Chang Heng, 86
charcoal, 187
Charlemagne, 42, 43, 45, 48, 50–51,
    55, 56, 57, 59, 62, 64, 67, 68, 69, 70,
    77, 171, 313
Charles V, king of France, 215
Charles Martel, 42, 55
Chartres, 159

Chartres Cathedral, 195, 198, 199
Château Gaillard, 145
Chaucer, Geoffrey, 230, 231, 233
checks (in banking), 185
Cheng Huo, admiral, 283
Chhang-an (Thang capital), 89
Childe, Gordon, 14
Ch'in dynasty, 85, 86
China, 15, 18, 37, 46, 55, 69, 82–99,
    112, 114, 115, 145, 147, 151, 154,
    156, 157, 166, 167, 179, 182, 183,
    200, 206, 207, 208, 211, 218, 230,
    233–234, 241, 242, 245, 250, 266,
    271–272, 278, 283, 287, 288
Chrétien de Troyes, 123
chronometer, 279
churka (cotton gin), 99, 122
Cicero, 10, 243
Cipolla, Carlo, 41, 208, 211, 263, 275,
    281, 288, 290
Cistercian Order, 9, 112, 114, 117, 139,
    174, 290–291
Cîteaux, 114
cities, 37, 43, 102, 107–108, 123–124,
    186–192
  charters of, 108, 116
Cividale, siege of (1331), 208
Clairvaux, 114, 117, 188–190, 200
clepsydra, 11, 19, 22, 86, 91, 210
Clermont Cathedral, 67
climate, 43
clinker construction, 72, 75, 155
clock, mechanical, 3, 89–92, 210–215,
    273–274, 287, 289, 319
cloth making, 17, 19, 25, 37, 49–54,
    103, 114, 118–123, 124, 173–181,
    192, 205, 259–260, 262, 269–271,
    289
  application of water power to, 114,
    178, 288–289
  women's role in, 25, 49–50, 118, 120,
    173–181
clothing, 32, 103, 180, 181
Clovis, king of the Franks, 48
Cluny, abbey of, 129, 130, 132
coal, 187, 203
cod, 282
cofferdam, 217
cog, 75, 154–156, 157, 221–222, 277
Colleoni, Bartolomeo, 249
Cologne, 107, 186
Cologne Cathedral, 67
Columbus, Christopher, 234, 240, 246,
    255, 277, 279, 284, 286
Columella, 36, 170
commenda, 108–109
Commercial Revolution of the Middle
    Ages, 105, 107, 167–168, 169, 220,
    246, 281
communes, 108
company, commercial, 169, 183–186
compass, magnetic, 1, 3, 86, 93–95,
    157, 222–223, 224, 278
Compostela, 148
concrete, 26, 37, 67
conservation, 6, 171, 290–291
Constantinople, 40, 50, 61, 66, 107,
    123, 148, 179, 183, 237, 240,
    248–249, 263
  fall of (1453), 3, 237, 240, 248–249
Contamine, Philippe, 148
copyists, 78–79, 181–182, 190, 291
Corbie, 79
Cordova, 188
cotton, 20, 24, 50, 103, 122–123,
    179–181, 270–271
  German, 270–271
  Muslim, 50, 103, 122
  North Italian, 122–123, 179–181,
    270–271
Council of Florence (1441), 240
Council of Nantes (660), 49–50
Covilha, Pero da, 240
crafts, 8–9, 25, 79–80, 86, 87, 118–128,
    158, 173–183, 186, 191, 291, 316
  women in, 25, 49–50, 118, 125, 176,
    289
crank, 65–66, 82, 128, 255, 257
Crécy, battle of (1346), 204, 205
credit arrangements, 37, 108–109, 169,
    185–186
  see also cambium, commenda
Cremona, 180
crenellation, 142
Crete, 19
crop rotation, 102, 107–108, 112
  see also open field system
crops, 102, 103, 113, 173, 285
crossbow, 66, 147–148, 204, 247–248,
    251, 317
Ctesibius, 22, 61
culverin, 247–248
cupellation, 128, 266
curb bit, 57
curtain wall, 142
"customs of miners," 128–129

Dales, Richard C., 159, 164, 228, 289–290
Damascus, 181
Damme, 221
dams, 102, 114, 116–117
Daniel-Rops, Henri, 309
Dante, 211, 239
"Dark Ages," as a term, 2–3
Datini, Francesco, 184–185, 239–240
De rebus bellicis, 304
Della Robbia, Luca, 275
Della Torre, Marc Antonio, 258
demesne, 109–110, 111
Derry, T. K., 246
Dias, Bartholomeu, 240
Diaz, Bernal, 281
Dijksterhuis, E. J., 230
Dijon, 48
Dinant, 123
Diophantine algebra, 226
Dioscorides, 100
diseases, 286; see also Black Death
dissection, 229
distillation, 102, 163
Dockès, Pierre, 307
Domesday Book, 113, 117, 171, 313
domestic architecture, see architecture, domestic
Donatello, 275
donats, 241
Donatus, Aelius, 241
Dondi, Giovanni di, 214
Dondi, Jacopo di, 214
Dorestad, 43
Douai, 174–175
doublet, 181
Dover, 117
drawbridge, 143
drawplate, 125
drilling, deep, 85, 87, 112
dromon, 71
Duby, Georges, 3, 42, 113, 171, 172, 173, 269
Duomo (Cathedral) of Florence, 254, 255, 271
Durham Cathedral, 131–132, 135
dyeing, 17, 25, 50, 120–121, 122, 178, 180–181, 190
Dyer, Christopher, 188, 190

Eadwine, 79
edge runner mill, 89, 114
education, 36, 159, 160, 190, 316

Edward I, king of England, 147, 185
Edward II, king of England, 185
Edward III, king of England, 185–186, 208
Egypt, 19–20, 37, 82, 103, 167, 240, 288
Eilmer of Malmesbury, 238–239
emporia, 43
enclosure movement, 268–269
England, 47, 113, 120, 141, 178, 181, 187, 222, 268, 269, 282, 289, 313
see also Britain
engraving, 242, 245
equatorium, 211
Eratosthenes, 22
Eric the Red, 106–107
Ericsson, Leif, 106–107
escapement, 89–92, 199, 211–212, 214, 318
estate management, 169–172
Ethiopia, 240
Euclid, 22, 160, 227
Europeanization, 286–287
exchequer, English, 159
Exploration, Age of, 3, 279–284
motives for, 281–282
eyeglasses, 3, 227, 228

Fabriano, 182
factors, 183, 314–315
factory system, 269–270, 289
Falaise, 143
falconry, 233
Fall, Adam's, 6–7, 14
famine, 171–172
felting, 17
Ferghana, 85
feudalism, 80, 250
origins of, 14, 55, 250
Feuerbach, Georg, 257
"Fibonacci sequence," 226
Finley, M. I., 22, 37
fire lance, 207
fireplace, 144, 145
First Crusade, 102, 105, 140, 141
flail, 31
Flanders, 120, 122, 134, 173–175, 178, 185, 186, 191, 215, 222, 225
flax breaker, 179
flintlock, 247
Florence, 121, 178, 183, 184, 185–186, 207–208, 218, 239, 240, 254, 255, 262, 269, 270, 271, 275, 280, 314
Flower, C. T., 70

flyer spindle, 176, 269
flying buttress, 130, 134–135
flying machine, 238–239, 252, 262
food preservation, 19, 37, 279–280
foods, introduction of, 102–103,
    279–280, 284
Forbes, R. J., 76, 200, 230
force pump, 22, 61
forecarriage, movable, 219
forests, depletion of, 4, 156, 171, 172,
    290–291, 313
  regeneration of, 268–269
Fougères, 142
Fourth Crusade, 179
Fra Angelico, 275
Fra Lippo Lippi, 275
Francis of Assisi, Saint, 233
Frankfurt, 246
Frederick II, Holy Roman Emperor,
    230, 233
Frisians, 51, 75, 112
Fugger, Jacob, 266
Fugger family, 270
Fulk Nerra, 140
fulling, 17, 25, 114, 120, 178
Funcken, Johannes, 266
furlongs, 111
fusee, 271, 272
Fust, Johann, 241, 245
fustian, 270, 271

Gaddi, Taddeo, 218
Gale, W. K. V., 63, 267
Galen, 100, 160, 289
galleys, 29, 72–74, 156, 158, 222
Gama, Vasco da, 283
garbage disposal, 190
Gattamelata, Erasmo, 275
Gaul, 31, 41, 43, 67, 69
Gawain, 123
gears, 21–22, 34–35, 37, 114, 258
Genoa, 105, 107, 156, 158, 167, 168,
    183, 184, 186, 222, 224, 225, 234,
    240, 277, 280, 283, 284
geography, 101, 161–162, 234, 249
geomancy, 86, 94
geometry, 13, 136, 160, 194, 197–198,
    199, 200
  in architecture, 136, 194, 197–198,
    199, 200, 316
George, Nadine, 128
Gerard of Cremona, 160
Gerbert (Pope Sylvester II), 159–160

Germany, 41, 135, 168, 181, 184, 240,
    241, 242, 243, 268–269, 270, 271,
    289
Gervase of Canterbury, 136–139, 192
Ghent, 186, 208
Gibbon, Edward, 1, 2, 4, 15–16, 31, 39,
    80
Gibraltar, Strait of, 158, 221, 234, 282
Gille, Bertrand, 3, 18, 205, 251, 255,
    256, 257, 258, 262, 289, 290
Gimpel, Jean, 4
Giotto, 239, 240
Gisors, 141
Glanvill, Joseph, 16
glass, 25, 37, 43, 102, 123, 128, 129,
    132–134, 136
  stained, 68, 129, 132–134, 136
Gokstad ship, 73–74
gold, 163, 168, 281
goldsmiths, 124, 215
Gothic, see architecture, Gothic
Goths, 70, 281
Gozzoli, Benozzo, 275
Grand Banks, 282
Grand Canal of China, 86
Grand Pont, Paris, 71, 152, 153
Great Galley, 221
Great Schism, 235
Great Wall of China, 85–86, 234
Greece, 19–20, 37, 160, 286, 287, 288,
    296
  see also Byzantium
Greek fire, 60–62, 71, 206
Greek science, 21, 23, 36. 77, 100–101,
    159–160, 228–229, 237
Greenland, 106, 154, 283, 306
Gregory I (the Great), pope, 130, 132
Gregory III, pope, 47
Gregory of Tours, 48, 49, 67
Grosseteste, Robert, 227, 230, 246
guilds, 107, 121, 124, 129, 176, 190,
    199
  attitudes toward innovation of,
    124–125, 176
  craft, 124–125, 129, 176, 199
  merchant, 197, 121, 176
Guillaume de St. Cloud, 231
gun carriage, 248, 249
gunpowder, 1, 61, 95, 204, 205–207,
    208–209, 247
gunpowder firearms, 3, 95, 204, 206–210,
    237–238, 240, 247–252, 287
  social effects of, 249–252, 287

guns, hand, 209, 247–248, 249–250
Gutenberg, Johann, 241, 242–245, 274
Gutenberg Bible, 243, 244–245
gynaeceum, 49–50, 118

Habsburgs, 240, 266
hackling, 54
Halesowen, 172
Hall, Bert S., 238, 252–253, 318, 321
Hall, Rupert, 267
Hallstatt civilization, 82
Hamburg, 107
Hamlet, 193
Hamwih, 43
Han dynasty, 93–94
Hanseatic League, 107, 156, 168, 222, 240
harbor installations, 30–31, 154, 156, 275
harness, 3, 14, 20, 32, 45–46, 47, 49, 107–108, 149, 219
harrow, 45, 46
Hart, Ivor, 258
Harun-al-Rashid, 50–51, 100
harvester, 31
Harvey, John, 125
Haskins, Charles Homer, 77, 235, 237
heddle, 53–54, 119–120
hemp, 114
Henry I, king of England, 149
Henry II, king of England, 141
Henry III, king of England, 230
Henry the Navigator, 282–283
Herculaneum, 24
Herjulfson, Bjarni, 106–107
Heron of Alexandria, 22, 93, 252
herring, 76, 225
Hill, Donald, 210
Hindu-Arabic numerals, 3, 99, 101, 225–227, 246
Hippocrates, 100
hoisting machinery, 194–196, 199, 271
Homer, 18, 19, 20, 21, 100
homo ludens, 252
Honorius of Autun, 11
hops, 125
horse, 20, 46–47, 58, 149
    domestication of, 20
    in traction, 46–47, 107–108, 149
horse collar, 45–46, 47, 81, 82, 149
horseshoe, nailed, 46, 56–57, 149, 250
Hospitalers, 141, 143
Housman, A. E., 252

Hrabanus Maurus, 77, 133
Hugh, abbot of Cluny, 132
Hugh of St. Victor, 12–13, 288
Hugo, Victor, 309
Huizinga, Johann, 253
Hunayn ibn-Ishaq, 100
Hundred Years War, 3, 168, 185–186, 208, 235, 248
Hungary, 219, 266
hurdy-gurdy, 66
Hussite War Ms., author of the, 255
Hussite wars, 217
hypocaust, 27

iatrochemistry, 163
Ibn Jubayr, 158
Ibn Sa'id, 312
Iceland, 106, 155, 283, 306
al-Idrisi, 161–162
Ile-de-France, 216
incunabula, 245–246
India, 15, 18, 27, 37, 55, 82, 99, 102, 104, 116, 131, 175, 218, 226, 240, 255, 281, 287
indigo, 181
Indo-China, 97
infield-outfield system, 112
ink, 78, 240, 241, 243–244, 274
Inuit, 82, 283
invention, process of, 2, 238, 241, 253, 254
Ipswich, 43
iron, 19, 57, 62–65, 80–81, 86, 87–88, 107, 114, 200–203, 266–267
Iron Age, 17, 18, 19
irrigation, 11, 19, 23, 40, 102, 103, 113, 114
Isidore of Seville, 11, 77, 246
Islam, 42, 100–103
ivory, 283

James of St. George, master mason, 196
Janszoon, Laurens, 242
Japan, 89, 97
Jean de Colmieu, 59
Jenson, Nicolas, 243
Jerome, Saint, 8
Joan of Arc, 247
John, king of England, 140
John II, king of Portugal, 249, 278
John of Brokehampton, abbot, 171
John of Pian de Carpine, 166
John Scotus Erigena, 11, 77

Jumièges, abbey of, 67
Jundi-Shapur, 100
junk, Chinese, 99, 283

Kapellbrücke, 218
Karl (Charles) IV, Holy Roman
    Emperor, 217
Karlsbrücke, 217
Khaifeng, 85, 89, 90
al-Khwarizmi, 160, 191, 225
Kilwardby, Robert, 13
knights, 43, 55, 58, 249–250, 251–252
knitting, 179
knorr, 75–76, 155
Kocs, 219
Korea, 87, 89, 97, 99, 241, 242
Krak des Chevaliers, 143–144
Kranzberg, Melvin, 291, 327
Kyeser, Konrad, 250–251, 252, 255, 261

labor, attitudes toward, 7–14
Labrador, 106
Lambert of Ardres, 60
land clearance and drainage, 6, 47–48,
    111, 113, 164
Langeais, 140
Langres, 70
Lapphyttten, 200
lateen sail, 3, 30, 71, 81, 158, 221, 222,
    277, 297
Lateran Council of 1139, 148
lathe, 64–65, 128, 199, 259
latifundium, 36, 44
latitude, determination of, 231, 278
Laurence Vitrearius, 134
law, profession and study of, 160
law codes, Germanic, 48, 50
Lefebvre des Noëttes, Richard, 14, 32
Le Goff, Jacques, 3, 214, 215
Lei Tsu, 83
Leland, John, 270
Le Mans, 69
lens, 227, 289
Leonardo da Vinci, 93, 238, 239, 247,
    252–253, 254, 255, 256, 257,
    258–263, 264, 265, 267, 268, 269,
    271, 272, 273, 274, 275, 290, 321,
    324–325
Leonardo Fibonacci (Leonardo Pisano),
    225–226, 233, 246, 257, 320
letter of credit, 186
Li Kao, 86
Liber abaci, 224, 320

liberal arts, 10–11
Liège, 201, 267
lighthouses, 30–31, 69
Lillers, 112
linen, 20, 50, 51, 54, 122, 179, 271
linseed oil, 243, 274
Lisbon, 282, 283
Livre des Mestiers of Bruges, 176
llama, 285
Loire, 216–217, 247
London, 43, 107, 183, 186, 188, 191,
    202, 220, 245
London Bridge, 151–152, 153
longa caretta, 219
longbow, 204, 205, 317
longship, 29, 73–75, 156
loom, 17, 19, 25, 52–54, 86, 118–120,
    175, 177–178, 180
    horizontal, 85, 118–120, 177–178
    tapestry, 178
    vertical two-beam, 54, 178
    vertical warp-weighted, 52–54
Lopez, Robert S., 3, 105, 121, 124, 169,
    178, 216, 237, 280
Lorraine, 134
Louis IX, king of France, 166
Lucca, 123, 178, 179, 185
Lucerne, 218
lute, 102
Lyons, 70, 151

Ma Chun, 86
machicolations, 142
Madeira Islands, 283
Magna Carta, 141
magnetic needle, 94, 157
Mahomet II, 248–249
mainspring, 273
Mainz, 241, 242
maize, 285
Malocello, Lancelotto, 234
al-Mamun, caliph, 100
mangonel, 147
manor, 109–110
manorial accounts, 170
manorial system, 14, 109
map-making, 161–162, 223–224, 234
Maricourt, Pierre de (Petrus Pere-
    grinus), 231, 233
Marlenheim, 49
Marne, 70
Marseilles, 125, 158, 277
Martianus Capella, 77, 246

Martini, Francesco di Giorgio, 257, 262, 263, 272, 288
master mason, 5, 136–139, 192–200, 218, 252, 316
  God as a, 5, 192
matchlock, 247–248
mathematics, 13, 37, 225–227, 233, 257
Maximilian, Holy Roman Emperor, 249, 252, 266
mechanical arts, 10–14, 236
medical schools, 159, 229
Medici, Cosimo de', 275
Medici family, 169, 275
medicine, 159, 160, 230–231, 258
Mesopotamia, 17–19, 82
Messina, 158
metallurgy, 18, 19, 24, 32, 37, 40, 62–65, 80, 86, 99, 114–115, 128, 200–203, 265, 266–267, 268
metalwork, 128
meurtrière, 142, 143
Middelburg, 275
Middle Ages, perceptions of, 1–4
  dating of, 2–3, 41–42
Milan, 123, 180, 184, 213, 240, 252, 270, 273
military organization, 31, 252
military technology, 31, 55–62, 102, 139–148, 247–252
mills, 11, 23–24, 34–35, 48–49, 85, 87–89, 113–114, 153, 252, 307
  animal-powered, 23–24, 35, 115
  floating, 70–71, 87, 89, 113, 153
  numbers of, 113–115
  tidal, 117
  water-powered, 11, 34–35, 48–49, 85, 87–89, 113–114, 252
  wind-powered, 99, 117
  see also waterwheel
Ming dynasty, 91
mining, 24, 40, 62, 128–129, 168, 240, 265, 266, 267, 286
Mokyr, Joel, 100
Moldau, 217
Mongols, 91, 92, 99, 166–167
Monkwearmouth, 68
monochord, 77
Monte Cassino, 131–132
Monte Settimo, 216
Montreuil, Pierre, 193
mortar, 26, 67
movable type, see printing
Muhammad, 102, 103

Müller, Johannes (Regiomontanus), 256–257
multure, 49, 115–116
Mumford, Lewis, 4
Munro, John, 122
Murano, 123
Murcia, 116–117

national states, 249–250, 290
nature, attitudes toward, 6–7, 8, 233
navigation, 20–21, 30, 69, 71–76, 94, 157, 158, 221–225, 240, 275–279, 281–287, 326
Neckam, Alexander, 119–120, 157
Needham, Joseph, 15, 50, 83, 85, 87, 91, 115, 206, 265, 287
Netherlands, 43, 47–48, 76, 112, 181, 221, 241, 242, 264, 275
new cities, 107–108
Newfoundland, 106, 306
Nicholas of Cusa, 255
Nicolas de Biard, 192
Nina, 277, 278
noria, 34
Normandy, 134, 136
Notker, 57, 62
Notre Dame de Paris, 129, 135, 154, 199
Nuremberg, 123
Nydam ship, 72

oats, 112
odometer, 93
oil painting, 243, 274
open-field system, see agriculture, open field
optics, 160, 227–228
Orban, 249
Oresme, Nicolas, 234–235, 255
organization of work, 109–112, 118–120, 125, 128–129, 269–271, 289
Orléans, 159, 217
  siege of, 247
Ostia, 24
Otto of Freising, 123
Ottoman Turks, 241, 248–249
Ovitt, George, Jr., 4, 6, 165
ox, 20, 45, 46–47
Oxford University, 160, 227, 230

Pacey, Arnold, 175, 194, 199
Pachomius, 8–9
Pacioli, Fra Luca, 257

paddlewheel boat, 86, 89, 264, 265, 304
Padua, 214
Palais-Royal, 215
Palermo, 123, 179
Palestine, 141
Palladio, Andrea, 218
Palladius, 8, 31
paper, 95–97, 127, 182, 240, 241
paper money, 96, 305
papyrus, 11, 20, 96, 97
parachute, 239
parchment, 21, 78, 96, 97
Paris, 71, 79, 113, 129, 145–146, 152,
    153, 154, 159, 183, 186, 215, 245,
    316
Paston family, 227
patent, 254, 325
Pavia, 180
  siege of, 57
paving, 19, 28, 29, 187, 188, 216
Peasants' Rebellion of 1381, 116,
    172–173
Pedro IV, king of Aragon, 225
Pegolotti, Francesco Balducci, 280–281
Peking, 91, 305
perpetual motion, 231–232
Persia, 84, 85, 87, 89, 99, 102, 183, 200
Pessagno, Manuel, 167
Peter of Colechurch, 151
Petrarch, 208
petroleum, 61
Pharos of Alexandria, 31
Philip VI, king of France, 205
Phillips, J. R. S., 282
Phoenicians, 20–21
Pi Sheng, 98
Piacenza, 180
Picardy, 113, 126, 198
pig iron, 202
pilgrims, 148, 156, 222
Pinta, 277, 278
pipe organ, 128
Pirenne, Henri, 39–40, 43
Pisa, 105, 186, 224, 225, 233
Pius II, pope, 279
pivoted axle, 149
plagiarism in invention, 262, 324–325
Plato, 21, 100, 245
Plautus, 255
Pliny, 31, 36, 44, 84, 281
plow, heavy, 3, 14, 43, 44–45, 46, 47,
    49, 111–112
  light (aratrum), 23, 45

Poitiers, battle of (732), 42
  battle of (1356), 204
pollution, 4, 186–187, 203, 313
Polo, Marco, 166, 181, 233–234, 279,
    305
Pompeii, 24, 25, 29
Pont d'Avignon, 150, 152, 153, 216
Pont du Gard, 26, 27
Pont Ecumant, 216
Pont-St.-Esprit, 151, 153, 218
Ponte Sant'Angelo, 28
Ponte Vecchio, 218
population, 4, 41, 42–43, 109, 111,
    112, 171–172, 186, 235
porcelain, 92
portolan charts, 223–224, 278
Portugal, 167, 181, 210, 234, 240, 277,
    278, 279, 282–284
postal service, see scarsella
Postan, Michael, 170
pottery, 18, 25, 37, 40, 43, 87, 92, 107,
    286
Prague, 217, 273
press, 19, 24, 37, 240, 241, 244–245
  beam, 19
  printing, 240, 244–245
  screw, 24, 37, 244–245
Price, Derek de Solla, 289, 296–297,
    326
printing, 1, 86–87, 97–99, 237, 240,
    242–246, 263, 287, 289, 309
  with movable type, 3, 87, 98–99,
    242–243
  wood-block, 98, 241–242, 243
printing shops, 245
progress, concepts of, 3–4, 287–288
Ptolemy, 22, 101, 160, 161–162, 230,
    257, 279, 282
putting-out system, 3, 173–175, 269,
    289
pyramids, 19
Pythagoras, 77

quadrant, 278
Quentovic, 43

rails, 266
Raymond, archbishop of Toledo, 160
rebec, 102
Reconquista, 167, 282
reduction furnace, 62–63, 80
reeve, 170
Regiomontanus, see Müller, Johannes

Renaissance, 237–238, 239
renaissance of the twelfth century, 158–164, 237
Reti, Ladislao, 321
retting, 54, 122
Reynolds, Robert, 80, 107, 123
Reynolds, Terry S., 288, 289
Rheims, 159
Rheims Cathedral, 67, *135*, 198
Rhône, 150
Riccardi company, 185
Richard I (Lionheart), 107, 145
Richard of Wallingford, 211, 227, 318
Richer, 160
ridge-and-furrow plowing, 112
"Ridley's stripper," 31
roads, 28, 69–70, 148–149, 216
    maintenance of, 69, 149, 216
    see also paving
Robert of Chester, 162
Robinson, Fred C., 3
Roc Amadour, 148
Rochester Castle, *140*
Roger II, king of Sicily, 123, 161–162
Roger of Hoveden, 158
Roger's Book, 161–162
rolling mill, 267
Roman de la rose, 211
Roman Empire, fall of the, 3, 39–40, 80
Rome, 17, 23–38, 70, 147, 188, 235, 286, 287, 288
Rosa Ventorum, 222–223
rotary grindstone, 65–66
rotary quern, 19
Rothair, Lombard king, 48
round ship, 29–30, 75–76
Russia, 75, 207

saddle, 55, 56, 250
sails, 20, 30, 71, 72–74, 99, 154–155, 158, 221–222, 277–278, 297
St. Albans, abbey of, 116
St. Denis, abbey of, 67, 132, 135, 136
    fair of, 107
St. Eustorgio, Milan, 213
St. Gall, abbey of, 56, 79–80, 114
St.-Germain-des-Prés, abbey of, 49
St. Gothard Pass, 216
St. Philibert, abbey of, 68
St. Riquier, abbey of, 79
Sainte-Chapelle, 193
Salerno, 163, 229
Salernus, 163

Samarkand, 84, 85, 97
Santa Maria, 277, 279, 285
Santa Sophia, 66
scarlets, 121–122
scarsella, 183
Schoeffer, Johann, 241–242
Schoeffer, Peter, 241, 242, 245
science, 13–14, 21–23, 36, 76–77, 82–83, 99, 100–101, 158–164, 225–235, 237, 273–274, 289–290, 296–297
    Church's attitude toward, 229–230
Scientific Revolution, 289–290
screw, 22, 37, 265
screw jack, 181, *198*, 199
Seneca, 36
serfs, 44, 108, 109, 111, 172–173
sewage systems, 19, 66, 186, 188, 315
shaduf, 19
Shakespeare, William, 172–173, 244, 258
sheep farming, 23, 103, 173–174, 178–179, 268
Shelby, Lon R., 316
shipbuilding, 20, 28–30, 37, 71–76, 99, 154, 221–225, 238, 275–278, 302
shock combat, 55, 56, 58
shuttle, 119–120
Sicily, 97, 102, 103, 105, 106, 107, 122, 158, 160, 161
    Norman conquest of, 102, 105, 122, 161–162
Siena, 254, 257
silk, 25, 50, 82, 83–85, 96, 97, 122–123, 178, 179, 270
Silk Road, 83, 84, 97
silk-throwing, 178
Singer, Charles, 227, 258
slavery, 24, 36, 44, 49, 80, 118, 124, 128, 286, 287
slitting mill, 266–267
Sluys, 76
smallpox, 286
soap, 31, 50, 125
Spain, 97, 103, 105, 106, 114, 116–117, 122, 140, 159, 168, 170, 181, 182, 234, 265, 266, 284, 286
Speyer, 130
spice trade, 279–281, 283, 284
spinning, 17, 25, 49, 51–52, 122, 175–176, 260, 269, 270–271
spinning wheel, *175*–176, 179, 180, 269, 286

spinning wheel (cont.)
flyer, 176, 269
treadle, 176, 269
Spreuerbrücke, 218
spur, 57
standards of living, 41, 191, 286–287
Statute of Laborers, 172
steel, 32, 99
steering oar, 20, 30, 157–158, 223, 309
Stephenson, Carl, 109, 160
stern rudder, 99, 156, 157–158, 277, 309
Stiefel, Tina, 160, 164
still, 163
stimulus diffusion of technology, 15, 87, 91, 99
Stirling, siege of (1304), 147
stirrup, 14, 55–56, 60, 82, 149, 250
origin of, 55–56
social effects of, 55, 250
stock farming, 103, 171, 172, 173
stomacher, 181
Stone Age, 17–18, 37, 82
Stradanus, Johannes, 2
Strasbourg, 242
Strasbourg Cathedral, 200, 214
Strato, 21
strikes, 175
Styria, 201
Subiaco, 243
Suger, abbot of St. Denis, 132, 136
Sumer, 20
Su Sung, 89–92, 211
Sung dynasty, Northern, 85, 87, 92
Sutton Hoo ship, 72, 73
Switzerland, 216, 218
swords, 64
Synod of Arras (1024), 130
syphilis, 286
Syria, 25, 102, 103, 141, 222, 280

Taccola, Mariano di Jacopo, 254–255, 264
Tacuinum Sanitatis, 280
Tafur, Pero, 271
Talas River, battle of (751), 84, 97
tanks, 205, 206, 260
tanning, 17–18, 124, 188, 190
tapestry, see loom, tapestry
Tartars, Chin, 91
technological revolutions, 14–15, 17–20, 49
technology, attitudes toward

classical, 10, 36
medieval, 4–14, 288, 307
technology, transmission of, 15, 50, 82–85, 87–100, 102–104, 106, 167, 208, 287–288, 312, 327
Templars, 141
Tertullian, 6
textiles, 12, 50–51, 103, 107, 179–181, 270–271
Thames, 152
Thang dynasty, 86, 89
Theophanes, 61
Theophilus Presbyter, 127–128, 129, 133, 266, 274
Thomas Aquinas, Saint, 7, 229, 246
Thurzo, Jacob, 266
Tibet, 89
tidal mills, see mills, tidal
tilt hammer, 89, 266, 267
tiraz factories, 103
tires, 219
toilet paper, 96
Toledan Tables, 227
Toledo, 106, 160
tools, 17, 24–25, 37, 64–65, 124, 126, 128, 202, 268, 285–286
Toscanelli, Paolo, 254, 255, 282
Toulouse, 117
Tours, basilica of, 67
trade fairs, 108
trade partnerships, 43
translation, 36, 100–101, 160
transport, 28–29, 107, 154, 156, 186, 196, 218–220
cost of, 191, 219–220
speed of, 220
treadmill, 86, 89, 194–195, 199, 265, 266, 271
trebuchet, 145–147, 210
Tree of Battles, 251–252
Treix, 70
Trento, 129
Treviso Arithmetic, 257
Trezzo, 217
Trier Apocalypse, 46
trigonometry, 160, 225, 227
trip hammer, 88–89, 115, 200–201, 259
truss, see bridges, truss
Tu Shih, 86, 88
Tu Yu, 84
turbine, 257, 263–264
Turgot, Anne-Robert Jacques, 2
typefaces, 78, 243, 246

Unger, Richard, 42, 156, 275
universities, 160, 164, 227–229, 234
University of Bologna, 160, 229
University of Padua, 258
University of Paris, 160, 229, 230, 231, 234
Ursus, abbot of Loches, 48
Usher, Abbott Payson, 211
usury, Church's attitudes toward, 108–109, 185
Utrecht Psalter, 54, 56, 65–66

Valturio, Roberto, 256, 261
Van der Weyden, Rogier, 275
Van Eyck brothers, 274
Varro, 170
vault, 26, 68, 130–132, 139
    barrel, 26, 68, 130
    cross-rib, 130–132, 139
    groin, 130
Venetian Arsenal, 156, 158, 221, 271, 289
Venice, 102, 107, 117, 123, 125, 156, 158, 181, 222, 224, 240, 243, 245, 254, 257, 270, 271, 274, 280, 282, 291
Venta, Ugo, 167
verge-and-foliot, 211, 212–213
Vergil, Polydore, 257–258
Verrocchio, Andrea del, 258
Vigevano, Guido da, 205, 252, 256, 261
Vikings, 42, 59, 70, 74–76, 80, 105, 106, 140, 145, 281, 283, 306
Villard de Honnecourt, 135, 197–199, 205, 211, 217, 218, 231–232, 252, 253, 257, 318
Vincent of Beauvais, 13
Vire, 145
Vitruvius, 34–35, 36, 93
Vitry, Jacques de, 192, 196

wagons, 218–220
Walsingham, Thomas, 116
Walter of Henley, 170–171
Walterus, Landgrave of Hesse, 273–274
Wang Chen, 98
warper, 177, 180
watch, pocket, 273
water power, see mills, waterwheel
water-powered saw, 198, 199, 257, 259
water supply, 188–190
waterwheel, 33–35, 37, 48–49, 80, 81,

87–91, 97, 113–115, 164, 168, 178, 188–190, 199, 200–202, 265–266, 286, 288–289
    efficiency of, 115
    horizontal, 33, 49, 87–88, 115
    uses of, 35, 37, 87–91, 97, 114–115, 178, 188–190, 199, 200–202, 265–266, 288–289
    vertical, 34, 35, 49, 87–88, 113–115, 168, 265
    see also mills
weaver, 118–120, 173, 174, 175, 270–271
weaving, 17, 19, 25, 49, 52–54, 118–120, 122, 270–271
Weber, Max, 4
weir, navigation, 221
Wharram Percy, 110
wheel, 17, 37, 286
wheel lock, 247
wheelbarrow, 15, 92, 168
whippletree, 149
White, Lynn, Jr., 3, 5, 6, 14, 36, 40–41, 55, 65, 79, 80, 233, 307, 312
Whitney, Eli, 122
William I (the Conqueror), 113
William de Percelay, 220
William of Malmesbury, 238–239
William of Rubruck, 166, 206, 305
William of Sens, 136–139, 192
Williams, Trevor I., 246
Willibald, Saint, 76
Winchcomb, John (Jack of Newbury), 270
Winchester, 191
windmill, 99, 117, 118, 164, 264–265
    horizontal, 99
    tower, 264–265
    vertical, 117, 118
Windsor Castle, 196
wine, 19, 24, 43, 48, 69
wool, 23, 25, 43, 49–54, 103, 107, 120–122, 173–179, 183, 270

xylography, see printing, wood-block

yellow fever, 286
Yen Su, 86
Yuwen Khai, 86

al-Zarqali, 227
Zelochovice, 62

COPYRIGHT ACKNOWLEDGMENTS

CPSIA information can be obtained
at www.ICGtesting.com
Printed in the USA
LVHW042226210821
695831LV00021B/2262